Atlantis Series in Dynamical Systems

Volume 2

For further volumes:
http://www.springer.com/series/11155

Jaap Eldering

Normally Hyperbolic Invariant Manifolds

The Noncompact Case

ATLANTIS PRESS

Jaap Eldering
Department of Mathematics
Utrecht University
Utrecht
The Netherlands

ISSN 2213-3526
ISBN 978-94-6239-042-3 ISBN 978-94-6239-003-4 (eBook)
DOI 10.2991/978-94-6239-003-4

Softcover re-print of the Hardcover 1st edition 2013

Printed on acid-free paper

Series Information

The "Atlantis Studies in Dynamical Systems" publishes monographs in the area of dynamical systems, written by leading experts in the field and useful for both students and researchers.

Books with a theoretical nature will be published alongside books emphasizing applications.

Series Editors

Henk Broer
University of Groningen
Johann Bernoulli Instituite for Mathematics and Computer Science
Groningen, The Netherlands

Boris Hasselblatt
Department of Mathematics, Tufts University
Medford, USA

Atlantis Press
29, avenue Laumière
75019 Paris, France
For more information on this series and our other book series, please visit our website www.atlantis-press.com

Preface

In this work, we prove the persistence of normally hyperbolic invariant manifolds. This result is well known when the invariant manifold is compact; we extend this to a setting where the invariant manifold as well as the ambient space are allowed to be noncompact manifolds. The ambient space is assumed to be a Riemannian manifold of bounded geometry.

Normally hyperbolic invariant manifolds (NHIMs) are a generalization of hyperbolic fixed points. Many of the concepts, results, and proofs for hyperbolic fixed points carry over to NHIMs. Two important properties that generalize to NHIMs are persistence of the invariant manifold and existence of stable and unstable manifolds.

We shall focus on the first property. Persistence of a hyperbolic fixed point follows as a straightforward application of the implicit function theorem. For a NHIM the situation is significantly more subtle, although the basic idea is the same. In the case of a hyperbolic fixed point, we only have stable and unstable directions. When we consider a NHIM, there is a third direction, tangent to the manifold itself. The dynamics in the tangential directions is assumed to be dominated by the stable and unstable directions in terms of the respective Lyapunov exponents. Thus the dynamics on the invariant manifold is approximately neutral and the dynamics in the normal directions is hyperbolic; hence the name *normally hyperbolic*. The system is called r-normally hyperbolic, if the *spectral gap condition* holds that the tangential dynamics is dominated by a factor $r \geq 1$. An r-NHIM persists under C^1 small perturbations of the system. The persistent manifold will be C^r if the system is, but it may not be more smooth, even if the system is C^∞ or analytic. This can also be formulated as follows: r-normal hyperbolicity is an 'open property' in the space of C^r systems under the C^1 topology. The description above shows that the spectral properties of NHIMs and center manifolds are similar. The difference is that NHIMs are globally uniquely defined, while center manifolds are not.

There are two basic methods of proof for hyperbolic fixed points and center manifolds: Hadamard's graph transform and Perron's variation of constants integral method. Both can be extended to prove persistence of NHIMs, as well as existence of its stable and unstable manifolds. We employ the Perron method.

Both methods of proof construct a contraction scheme to find the persistent NHIM (and a similar contraction scheme can be used to find its stable and unstable manifolds). *Heuristically*, we can construct the implicit function $F(M, v) = \Phi^t(M) - M = 0$, where M is the NHIM and Φ^t is the flow of the vector field v after some fixed time t. Normal hyperbolicity of M implies that $D_1 F$ is invertible. Hence, there is a function $\tilde{M} = G(\tilde{v})$ that maps perturbed vector fields $\tilde{v}$ to persistent manifolds $\tilde{M}$, at least in a neighborhood of v. This idea does not work directly for higher derivatives. An inductive scheme can be set up that typically uses some form of the fiber contraction theorem. This scheme will break down after r iterations, hence the limited smoothness. Example 1.1 shows that this is an intrinsic problem.

To tackle the noncompact case, we replace compactness by uniformity conditions. These include uniform continuity and global boundedness of the vector field and the invariant manifold and their derivatives up to order r. We require additional uniformity conditions on the ambient manifold, namely 'bounded geometry'. This means that the Riemannian curvature is globally bounded, and as a result we have a uniform atlas which allows us to retain uniform estimates throughout all constructions in the proof.

Organization of the Book

This book is organized as follows. In the introduction, we give a broad overview of the theory of NHIMs with references to more details in the later chapters. We start by describing how NHIMs are related to hyperbolic fixed points and center manifolds. Then we give some basic examples and motivation for studying the noncompact case. We give a brief overview of the history and literature and compare the two methods of proof in the basic setting of a hyperbolic fixed point. Then we continue to introduce the concept of bounded geometry and a precise statement of the main result of this work and discuss its relation to the literature. We describe a few extensions and details of the results and conclude the chapter with notation used throughout this book.

Chapter 2 treats Riemannian manifolds of bounded geometry. We first introduce the definition of bounded geometry and some basic implications. We explicitly work out the relation between curvature and holonomy in Sect. 2.2. This we use in Sect. 3.7 to prove the smoothness of the persistent manifold. In the subsequent sections, we develop the theory required to prove persistence of noncompact NHIMs in general ambient manifolds of bounded geometry. We extend results for submanifolds to uniform versions in bounded geometry, to finally show how to reduce the main theorem to a setting in a trivial bundle. A number of these results are new and may be of independent interest, namely the uniform tubular neighborhood theorem, the uniform smooth approximation of a submanifold, and a uniform embedding into a trivial bundle.

In Chap. 3, we finally prove the main result in the trivial bundle setting. We first state both this and the general version of the main theorem and discuss these in full detail. We include a precise comparison with results in the literature, followed by an outline of the proof. Section 3.3 contains a discussion of the differences to the compact case and presents detailed examples to illustrate these. Then we start the actual proof. We first prepare the system: we put it in a suitable form and obtain estimates for the perturbed system. Then we prove that there exists a unique persistent invariant manifold and that it is Lipschitz. Second, we set up an elaborate scheme in Sect. 3.7 to prove that this manifold is C^r smooth by induction over the smoothness degree.

In Chap. 4, we discuss how the main result can be extended in a number of different ways that may specifically be useful for applications. We show how time and parameter dependence can be added and we present a slightly more general definition of overflow invariance that might be applicable to systems that are not overflowing invariant under the standard definition.

Finally, the appendices contain technical and reference material. These are referenced from the main text where appropriate. Appendix A shows an important idea that permeates this work: the implicit function theorem allows for explicit estimates in terms of the input, hence it 'preserves uniformity estimates'. This can then directly be applied to dependence of a flow on the vector field. In Appendix B, the Nemytskii operator is introduced as a technique to prove continuity of post-composition with a function. This is an essential basic part in the smoothness proof, together with the results on the exponential growth behavior of higher derivatives of flows in Appendix C. Here, we also develop a framework to work with higher derivatives on Riemannian manifolds. The last appendices include the fiber contraction theorem of Hirsch and Pugh that is used in the smoothness proof, Alekseev's nonlinear variation of constants integral defined on manifolds, and a brief overview of those parts of Riemannian geometry that we use.

What is (Not) New

Normal hyperbolicity can nowadays be called a classical subject; it was first formulated and studied in the late 1960s and 1970s, see the historical overview in Sect. 1.3. Although initial results were formulated for compact NHIMs, more recent work by Sakamoto and especially Bates, Lu, and Zeng have brought this to the noncompact setting, and even to semi-flows in Banach spaces.

The specific aspect that is new in this work is the differential geometric context in which our results on noncompact NHIMs are formulated. Uniformity of the ambient space seems not to have been addressed before in the literature, and our use of bounded geometry allows us to extend persistence of NHIMs to this setting, see Sect. 1.5 and Chap. 2. Additionally, some of our results on bounded geometry appear to be new, including the uniform tubular neighborhood theorem and the theorem on uniform smooth approximation of a submanifold.

The 'core' persistence proof itself is based on the Perron method; it can probably be replaced by the proof of Bates, Lu, and Zeng, that is based on the graph transform, when taking into account the necessary bounded geometry technical details from Chap. 2. Our proof uses ideas of Henry and Vanderbauwhede and Van Gils to extend the Perron method to NHIMs and higher smoothness. A novel aspect is that we develop these ideas on a trivial bundle with a bounded geometry manifold as base. This requires a whole framework to be set up, including representations of higher jets of a flow in Appendix C and formal tangent bundles of spaces of curves (see Sect. 3.7.4) to study derivatives of the Perron contraction operator. These ideas might be of interest in other contexts of dynamics in noncompact differential geometry.

Finally, our Definition 4.4 of a priori overflowing invariance might be new (although probably not surprising to experts in the field) and could prove useful for certain applications where the original definition of over- or inflowing invariance does not hold.

Acknowledgments

This book is based on my Ph.D. thesis work completed at Utrecht University. I am grateful to having had Hans Duistermaat as my supervisor while working on this project. His inspiring enthusiasm and deep insights have been greatly beneficial, and I feel honored to having been his Ph.D. student. After Hans' untimely death in 2010, the guidance of Erik van den Ban and Heinz Hanßmann has been instrumental in completing this project. I am indebted to both, and it was a pleasure working together.

The members of my thesis committee have provided valuable comments, and I would especially like to thank Charles Pugh for stimulating discussions and finding an error in a lemma.

Finally, I would like to thank the editors Henk Broer and Boris Hasselblatt for their support in preparing this book.

March 2013 Jaap Eldering

Contents

Chapter 1
Introduction

The basics of the theory of hyperbolic dynamics date back to the beginning of the 20th century, and the general formulation of the theory of normally hyperbolic systems was stated around 1970. Since then, many people have extended the theory, and even more people have applied it to problems in all kinds of areas.

Normally hyperbolic invariant manifolds are important fundamental objects in dynamical systems theory. They are useful in understanding global structures and can also be used to simplify the description of the dynamics in, for example, slow-fast or singularly perturbed systems.

In this work, we are specifically interested in noncompact normally hyperbolic invariant manifolds. We extend classical results that were previously only formulated for compact manifolds. However, in many applications the manifold is not compact, so an extension of the theory to the general noncompact case allows one to attack these problems in their natural context. The main result of this work is an extension of the theorem on persistence of normally hyperbolic invariant manifolds to a general noncompact setting in Riemannian manifolds of bounded geometry type.

1.1 Normally Hyperbolic Invariant Manifolds

We should first point out that the theory of (normally) hyperbolic systems can be applied to both discrete and continuous dynamical systems. That is, if we have a dynamical system (T, X, Φ) with X a smooth manifold and $\Phi\colon T \times X \to X$ the evolution function, then the system[1] is called discrete if $T = \mathbb{Z}$ and continuous if $T = \mathbb{R}$. In the discrete case, one typically has a diffeomorphism $\varphi\colon X \to X$ and the full evolution function is defined as $\Phi(n, x) = \varphi^n(x)$, i.e. iterated application of φ. In the continuous case, the map Φ is called a flow. It is generated by a vector field $v \in \mathfrak{X}(X)$ and in that case the map $\Phi^t : X \to X$ is again a diffeomorphism for any $t \in \mathbb{R}$.

[1] For simplicity of presentation we ignore the facts that Φ may have a smaller domain of definition, or that it is a semi-flow or semi-cascade, only defined on $T \geq 0$.

J. Eldering, *Normally Hyperbolic Invariant Manifolds*, Atlantis Series in Dynamical Systems 2, DOI: 10.2991/978-94-6239-003-4_1,

The two cases can be related by fixing a $t \in \mathbb{R}$ in the continuous case and then view $\varphi = \Phi^t$ as generating a discrete system. The statements of definitions and results are (almost) identical if formulated in terms of the evolution function Φ. The methods of proof share this similarity and can be translated into each other. We shall adopt the continuous formulation in this work, and refer to the evolution parameter $t \in T = \mathbb{R}$ as time. Even though our system is defined in terms of a vector field v, we call $x \in X$ a fixed point of the system when $\Phi^t(x) = x$ for any $t \in \mathbb{R}$. This is equivalent to saying that $v(x) = 0$, i.e. that it is a critical point of v; we adhere to the former terminology to better preserve the analogy with discrete systems.

Before we proceed to explaining normally hyperbolic invariant manifolds, it should be pointed out that these are a generalization of hyperbolic fixed points. Many of the characteristic properties generalize as well, so we first sketch the basic picture for hyperbolic fixed points. Let x be a fixed point of a vector field, $v(x) = 0$; it is called hyperbolic if the derivative $\mathrm{D}v(x)$ has no eigenvalues with zero real part. This means that the eigenvalue spectrum splits into parts left and right of the imaginary axis, that is, the stable and unstable eigenvalues, but no neutral ones. The corresponding stable and unstable eigenspaces $E_\pm$ are both invariant under the linear flow of $\mathrm{D}v(x)$ and these spaces are characterized by the fact that solution curves on them converge exponentially fast towards the fixed point under forward or backward time evolution respectively. It is a well-known result that there are corresponding stable and unstable (local) manifolds, denoted W^S_{loc} and W^U_{loc} respectively, which are the nonlinear versions of these, see Fig. 1.1. This situation can be generalized to a normally hyperbolic invariant manifold by replacing the single fixed point by a 'fixed set of points', that is, a manifold which is, as a whole, invariant.

Let us start with a somewhat informal explanation of the concept of a normally hyperbolic invariant manifold, which we shall from now on often abbreviate as a

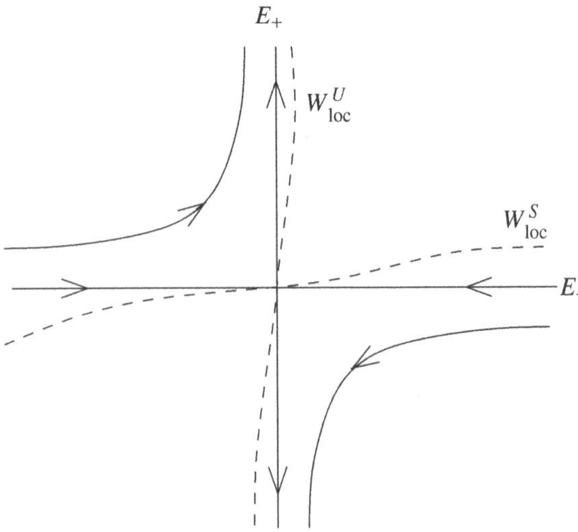

Fig. 1.1 A hyperbolic fixed point with (un)stable manifolds W^S_{loc}, W^U_{loc}

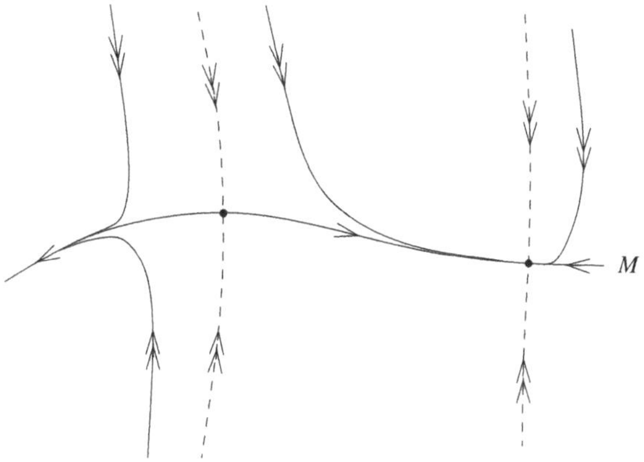

Fig. 1.2 A normally hyperbolic invariant manifold. The *single* and *double arrows* indicate slow and fast flow respectively

NHIM, as is common in the literature. If we have a dynamical system (T, Q, Φ) with phase space Q (which we shall often refer to as the 'ambient manifold') and evolution map Φ, then a manifold $M \subset Q$ is called invariant under the system if it is mapped to itself under evolution. In the continuous case this means that $\Phi^t(M) = M$ for all times $t \in \mathbb{R}$, that is, any point $x \in M$ stays in M, so its complete orbit is contained in M.

An invariant manifold M is then called normally hyperbolic if in the normal directions, transverse to M, the linearization of the flow Φ^t has a spectrum separate from the imaginary axis again. Although the precise definition is a bit more technical than in the case of a hyperbolic fixed point, the geometric idea is the same. The normal directions must separate into directions along which the linearized flow exponentially converges towards M and directions along which it exponentially expands; no neutral directions are allowed. Finally, the flow on M itself may expand or contract, but only at rates that are dominated by the expansion and contraction in the normal directions. Figure 1.2 shows part of a normally hyperbolic invariant manifold M that has only stable normal directions. Note that the dynamics on M itself can be very complex; it can have fixed points or even be chaotic. The only restriction is that the vertical contraction rate is stronger than horizontal ones (and similarly for expansion), as is indicated by the double and single arrows and visible from the convergence of solution curves to the rightmost fixed point on M.

1.1.1 Persistence and (Un)stable Manifolds

There are two important properties that generalize from hyperbolic fixed points to normally hyperbolic invariant manifolds. These are persistence of the fixed point and the existence of stable and unstable manifolds. The generalization of these properties is not a trivial statement nor easily proven in the generalized case of NHIMs, however.

Let us first focus on persistence. In case of a hyperbolic fixed point, this is trivially stated and proven. If the fixed point x is hyperbolic, then it will persist as a nearby fixed point under small perturbations of the vector field v and stay hyperbolic. The proof is a direct application of the implicit function theorem. If $\mathrm{D}v(x)$ has no eigenvalues on the imaginary axis, then certainly it has no zero eigenvalue, and therefore is a bijective linear map. So a slightly perturbed vector field $\tilde{v}$ will again have a fixed point $\tilde{x}$ nearby x and the eigenvalues of $\mathrm{D}\tilde{v}(\tilde{x})$ will be close to those of $\mathrm{D}v(x)$ if $\tilde{v} - v$ is small in C^1-norm. Hence the eigenvalues are still separated byAZ the imaginary axis. For a NHIM the situation is similar but technically much more involved due to the fact that there is no control on the behavior of solution curves in the invariant manifold. A normally hyperbolic manifold M does persist under C^1 small perturbations and the perturbed manifold $\tilde{M}$ is again normally hyperbolic and close to M in a precise way. The most important difference, however, is that $\tilde{M}$ generally has only limited smoothness, even if M and the system were smooth or analytic.[2] This smoothness is dictated by the *spectral gap condition*, which is roughly the ratio between the normal exponential expansion/contraction and the exponential expansion/contraction tangential to M. This fact already indicates that the proof of persistence of a NHIM cannot be a straightforward application of the implicit function theorem.

The stable and unstable manifolds generalize as well. That is, a normally hyperbolic invariant manifold M has stable and unstable manifolds $W^S(M)$ and $W^U(M)$ such that solution curves on these converge exponentially fast towards M in forward or backward time, respectively. Their intersection is precisely M. But there is actually more structure: these manifolds—we consider $W^S(M)$ but everything is equivalent for $W^U(M)$—are fibrations of families of stable and unstable fibers to each point $m \in M$,

$$W^S(M) = \bigcup_{m \in M} W^S(m).$$

We should be a bit careful with this last statement, as points $m \in M$ are generally not fixed points. These fibers $W^S(m)$ are invariant in the sense that the flow commutes with the fiber projection π^S:

$$\forall\, t \in \mathbb{R},\ m \in M,\ x \in W^S(m) \colon \pi^S \circ \Phi^t(x) = \Phi^t \circ \pi^S(x).$$

In other words, each fiber is mapped into another single fiber under the flow, namely the fiber over the flow-out of the base point m. This important fact means that if we use the fibration for local coordinates, then in these coordinates the horizontal, base flow decouples from the vertical, fiber flow. This is sometimes also called an *isochronous* fibration [Guc75] as all points in a fiber have the same long-term behavior. Each single fiber is as smooth as the system, but the dependence on the base point m, and thus the smoothness of the fibrations as a whole, is generally not better than continuous, see Fenichel [Fen74, Sect. I.G]. We do not investigate these invariant fibrations in the present work, although the mentioned results should hold for noncompact NHIMs as well.

---

[2] I do not know whether loss of smoothness is generic for NHIMs. See [Has94, HW99] for the case of Anosov systems.

1.1.2 The Relation to Center Manifolds

Normally hyperbolic invariant manifolds bear a close resemblance to center manifolds. Their spectral properties are roughly equivalent; they differ in the fact that NHIMs have an intrinsically global definition, while center manifolds are defined in local terms.

A center manifold $W^C_{\text{loc}}(x)$ of a fixed point x is a local invariant manifold such that its tangent space at the fixed point is the (generalized) eigenspace E_0 of the eigenvalues with real part zero, that is,

$$\mathrm{T}_x W^C_{\text{loc}}(x) = E_0. \tag{1.1}$$

We can extend the definition of center manifold a bit by including all eigenvalues λ with real part bounded by $|\mathfrak{Re}(\lambda)| \leq \rho_0$. An associated generalized center manifold consists of solutions that converge or diverge from x at an exponential rate bounded by ρ_0. Curves in the strongly[3] stable or unstable manifold converge or diverge at exponential rates larger than $\pm\rho_0$, respectively. These conditions can directly be compared to the description of NHIMs above, or Definition 1.8 (with $\rho_0 = \rho_M$).

If we take a look at Fig. 1.2 again, then we see that both fixed points (indicated with a dot) on M have (generalized) center manifolds; M itself is a center manifold for these, but for the rightmost fixed point we can actually construct the center manifold from any two solution curves converging to that fixed point from the left and right. For example, the union of the two curves drawn in the figure that converge to it could be taken as alternative center manifold. This reflects the well-known fact that center manifolds are generally not unique. This is the main difference with the case of NHIMs: center manifolds are only defined in terms of growth rates of solution curves locally with respect to one fixed point, while NHIMs are globally invariant objects, where the spectral splitting must hold everywhere along the invariant manifold. This difference is effectively the reason that center manifolds are not uniquely defined, while the perturbations of NHIMs are, see below. If we perturb the system in Fig. 1.2 a bit, then the persistent NHIM must everywhere be close to the original invariant manifold M. This enforces uniqueness; in Fig. 1.2 this is clearly visible: the alternative choice of center manifold to the rightmost fixed point diverges far from M. See also the example in Section 1.2.1.

There is a subtle question of smoothness both for center manifolds and NHIMs, related to the *spectral gap condition* (1.11). Center manifolds are arbitrarily smooth in a sufficiently small neighborhood of the fixed point x, but they are generally *not* C^∞, even though they satisfy an infinite spectral gap. See Van Strien's short note [Str79]. The reason is that the size of the neighborhood may depend on the degree of differentiability C^k. Persistent NHIMs generally have bounded smoothness due a finite spectral gap; but even if they have an infinite spectral gap, the smoothness

[3] We remove the eigenvalues associated to E_0 from $E_\pm$ so that E_-, E_0, E_+ together disjointly span the total tangent space at x.

of a persistent NHIM is (generally) not C^∞ for the same reasons. See Remark 1.12 and Example 1.3.

1.2 Examples

We present a few examples. The first detailed example serves to show explicitly that smoothness of a persistent manifold depends crucially on the spectral gap condition. The next examples motivate the usefulness of a noncompact version of the theory of normal hyperbolicity.

1.2.1 The Spectral Gap Condition

An invariant manifold is called an r-NHIM if the flow contracts or expands at exponential rates along the normal directions, and if these rates dominate any contraction or expansion along tangential directions at least by a factor r. This separation between growth rates along directions tangential and normal to the NHIM is encoded in Eqs. (1.10) and (1.11).

Here we introduce a simple example where the growth rates can be identified with eigenvalues λ of the linearization of the vector field at stationary points. Furthermore, we consider the simplified case where only a stable normal direction is present. That is, we consider a flow that contracts in the normal direction at an exponential rate of at least $\rho_Y < 0$ and along the invariant manifold it contracts at most at the rate ρ_X with the simplified spectral gap condition

$$\rho_Y < r\,\rho_X \qquad \text{with} \quad \rho_X \leq 0,\ r \geq 1. \tag{1.2}$$

The spectral gap is fundamental to persistence of invariant manifolds: the compact invariant manifolds that are persistent under any small perturbation are precisely those that are normally hyperbolic[4] [Mañ78]. Mañé only proved this inverse implication for 1-normal hyperbolicity, the question is still open for r-normal hyperbolicity with $r > 1$. A further property of normally hyperbolic invariant manifolds is that the differentiability of a slightly perturbed manifold depends not only on the smoothness of the original manifold and the perturbed vector fields, but also on the spectral gap. The spectral gap determines an upper bound $1 \leq r < \infty$ on the smoothness of the perturbed system, as r has to satisfy[5] (1.11). This condition stems from the fact that

[4] The definition of normal hyperbolicity in [Mañ78] is a bit more general than the definition in this paper. That definition only requires a growth ratio $r \geq 1$ along solution curves in the invariant manifold, and not as a ratio of global growth rates ρ_X, ρ_Y, see also Remark 1.10.

[5] The case $r = \infty$ would require $\rho_X > 0$; when $\rho_X = 0$, any finite order r can be obtained, but only for perturbations sufficiently small depending on r.

when the flow has exponential growth behavior $e^{\rho t}$, then higher order derivatives will generally have growth behavior $e^{k\rho t}$ and the interval inclusion $[k\rho, \rho] \subset (\rho_Y, \rho_X)$ is required to show existence and uniqueness of the k–th derivatives via a contraction. The optimal differentiability degree r can be extended to a real number by viewing α– Hölder continuity as a fractional differentiability degree. That means that the perturbed manifold can be shown to be $C^{k,\alpha}$ when $r = k + \alpha$ satisfies the spectral gap condition and the system is $C^{k,\alpha}$ to start with.

The following example shows that this result is sharp. We construct a very simple compact, normally hyperbolic invariant manifold, and then show that an arbitrarily small perturbation yields a unique perturbed $C^{k,\alpha}$ invariant manifold, where $r = k+\alpha$ satisfies $\rho_Y = r\rho_X$. This in fact precisely violates the spectral gap condition, since that requires a strict inequality. The example could be adapted to obtain a perturbed manifold with smoothness no better than $C^{r'}$ for some $r' < r$, cf. Example 1.3. A more qualitative exposition of this example can also be found in [Fen72, pp. 198–200] and [Hal69, p. 239, 251].

Example 1.1 (Optimal $C^{k,\alpha}$ smoothness of persistent manifolds) Let the horizontal space $X = S^1$ be the circle and the vertical space $Y = \mathbb{R}$. Take two points $x_- = 0$ and $x_+ = \pi$ in X and set the vector field v to zero at $(x_-, 0), (x_+, 0)$. We turn these stationary points into hyperbolic fixed points, with v linear in neighborhoods around them and $Dv(x_-, 0), Dv(x_+, 0)$ having eigenvalues $\lambda_- < 0 < \lambda_+$ along X, respectively, and one global eigenvalue $\lambda_Y < \lambda_-$ in the vertical direction along Y, i.e. $\dot{y} = v_y(x, y) = \lambda_Y y$, see Fig. 1.3. We extend the horizontal component v_x of the vector field v to the whole space $X \times Y$ in such a way that it is C^∞, independent of y, and has no critical points except for x_-, x_+. Hence, $M = X \times \{0\}$ is an invariant manifold for the flow Φ^t of v.

First, we check that M is normally hyperbolic. The long time behavior of any point $m \in M$ is governed by its approach of the stable fixed point $(x_-, 0)$, except for $m = (x_+, 0)$. For $m = (x_+, 0)$ we have $D\Phi^t(m) = e^{Dv(m)t}$, hence

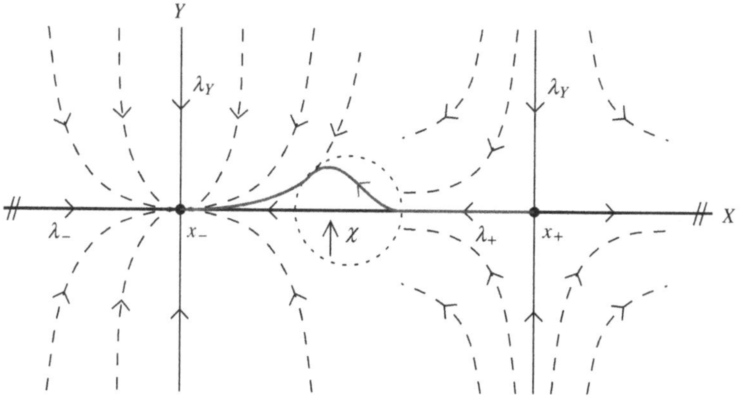

Fig. 1.3 An example invariant manifold exhibiting $C^{k,\alpha}$ smoothness under perturbation

$$\mathrm{D}\Phi^t(m)|_{\mathrm{T}_m X} = e^{\lambda_+ t},$$
$$\mathrm{D}\Phi^t(m)|_{\mathrm{T}_m Y} = e^{\lambda_Y t}.$$

More generally, consider a point $m \in M$ in the neighborhood of either $(x_\pm, 0)$ where v is linear. Then $\mathrm{D}\Phi^t(m)$ is given by

$$\mathrm{D}\Phi^t(m) = \begin{pmatrix} e^{\lambda_\pm t} & 0 \\ 0 & e^{\lambda_Y t} \end{pmatrix}$$

for as long as $\Phi^t(m)$ stays in that neighborhood of $(x_\pm, 0)$ where the vector field is linear. The transition time between these two neighborhoods is finite as v does not have zeros and the transition map preserves vertical lines $\{x\} \times Y$. The latter fact is because v_x is independent of y, that is, we have also found the invariant, foliated stable manifold of M. Gluing together these $\mathrm{D}\Phi^t$ maps on the different domains, we see that the resulting tangent flow splits again into independent horizontal and vertical parts, which can be estimated by

$$\forall t \leq 0\colon \left\|\mathrm{D}\Phi^t(m)|_{\mathrm{T}_m X}\right\| \leq C_X\, e^{\lambda_- t},$$
$$\forall t \geq 0\colon \left\|\mathrm{D}\Phi^t(m)|_{\mathrm{T}_m Y}\right\| \leq C_Y\, e^{\lambda_Y t},$$

where the constants $C_X,\ C_Y$ are determined by the flow Φ^t in the domain where v is nonlinear. For any point m close to $(x_-, 0)$ this estimate is sharp, hence we expect maximal smoothness $r = \lambda_Y/\lambda_-$ for a generic perturbation.

Next, we add a perturbation term $\varepsilon\,\chi$ to the vector field v, so we have a perturbed vector field $\tilde{v} = v + \varepsilon\,\chi$, where $\chi \in C_0^\infty$ is chosen with support on a small ball intersecting M away from the fixed points and pointing upward. This will 'lift' the invariant manifold as indicated in Fig. 1.3 for any $\varepsilon > 0$. Let $\tilde{M}$ denote this lifted manifold, that is, $\tilde{M}$ is the image of the two heteroclinic solution curves that run from $(x_+, 0)$ to $(x_-, 0)$ together with these fixed points. The solution curve that runs to the left is lifted up from the $x-$ axis after entering the region supp χ.

We first investigate two claims: that $\tilde{M}$ is invariant and that it is the unique invariant manifold that is close to M. The invariance is obvious; to the right of x_+ nothing has changed, so there $\tilde{M} = M$. To the left of x_+ we follow the original unstable manifold, get pushed up within the domain of support of χ and after leaving that domain and entering the linear flow around $(x_-, 0)$ we follow a standard curve ending at $(x_-, 0)$. This is a solution curve of $\tilde{v} \in C^\infty$, hence invariant and even smooth. Now assume there exists another invariant manifold M' nearby and let $(x', y') \in M' \setminus \tilde{M}$. The backward orbit of the point (x', y') must diverge to $|y| \gg 1$. If $x' = x_-$, then $y' \neq 0$ and this is clear. If $x' \neq x_-$, then the backward orbit will end up at a point (x, y) with x close to x_+ and $y \neq 0$; since we are in the linear domain of $(x_+, 0)$, this orbit will then diverge (in reverse time) along the stable manifold towards $|y| \gg 1$. Hence, M' is not close to M.

Next, we show that (for any $\varepsilon > 0$) the perturbed manifold $\tilde{M}$ is not more than $C^{k,\alpha}$ with $k + \alpha = \lambda_Y/\lambda_-$, even though the original and perturbed systems are

C^∞-smooth. To the left of $(x_-, 0)$, $\tilde{M}$ is given by the graph of the zero function from X to Y (as the continuation from $(x_+, 0)$ to the right along $X = S^1$). To the right of $(x_-, 0)$, the solution curve is given by $(x, y)(t) = (x_0\, e^{\lambda_- t}, y_0\, e^{\lambda_Y t})$, hence $y = C\, x^{\lambda_Y/\lambda_-}$ where C depends on x_0, y_0 only. So we can write $\tilde{M}$ as the graph of the function

$$\tilde{h}\colon X \to Y : x \mapsto \begin{cases} 0 & \text{if } x \leq 0, \\ C\, x^{\lambda_Y/\lambda_-} & \text{if } x > 0. \end{cases}$$

This function is exactly $C^{k,\alpha}$ for $k + \alpha = r$ in $x = 0$. Note that the loss of smoothness appears at a different place than the perturbation of the vector field. The relevant fact is that the different solution curves approaching the stable limit point have finite differentiability with respect to each other, and this depends on the horizontal and vertical rates of attraction at $(x_-, 0)$. ○

If we had assumed that $\rho_Y = \rho_X$, that is, $r = 1$, but with a non-strict inequality $\rho_Y \leq r\, \rho_X$, then normal hyperbolicity precisely fails and the invariant manifold indeed need not persist. By the arguments above it can already be seen that the persistent manifold can lose differentiability: when $r = 1$, the graph of the manifold will be given by

$$\tilde{h}(x) = \begin{cases} 0 & \text{if } x \leq 0, \\ C\, x & \text{if } x > 0, \end{cases}$$

which is clearly non-differentiable at $x = x_- = 0$. We can extend the example above to show that even more serious problems can occur.

Example 1.2 (*Non-persistence of non-NHIMs*) We consider Example 1.1 with $\rho_Y = \rho_X$. If we perturb the system with a small circular vector field around $x_- = (0, 0)$, then $Dv(0, 0)$ will have two eigenvalues $\lambda_Y \pm i\, \omega$ with $\lambda_Y < 0$ and $\omega \in \mathbb{R}$ small. Thus, the solution curves that should make up the invariant manifold around $(0, 0)$ will spiral in, which leads to the picture in Fig. 1.4. Note that the curves wind around the origin infinitely often. At the origin this is not a manifold anymore, and cannot be described by a function $\tilde{h}\colon X \to Y$. ○

The idea to perturb around the stable fixed point x_- also leads to the following example.

Example 1.3 (*Non-C^∞ persistence for $r = \infty$ NHIMs*) We consider again Example 1.1, but now with $\lambda_- = 0$. Then we have $\rho_X = 0$ and spectral gap $r = \infty$. If we let $\lambda_- = \varepsilon$ depend on the perturbation parameter $\varepsilon > 0$, then this decreases the spectral gap condition[6] to a finite number $r = \lambda_Y/\lambda_-$. Even though $r \to \infty$ as the perturbation size ε goes to zero, we still have a finite spectral gap for any fixed perturbation. We conclude that the corresponding perturbed manifolds are not C^∞,

---

[6] The ratio r in the spectral gap is defined by a strict inequality, which we ignore here for simplicity of presentation.

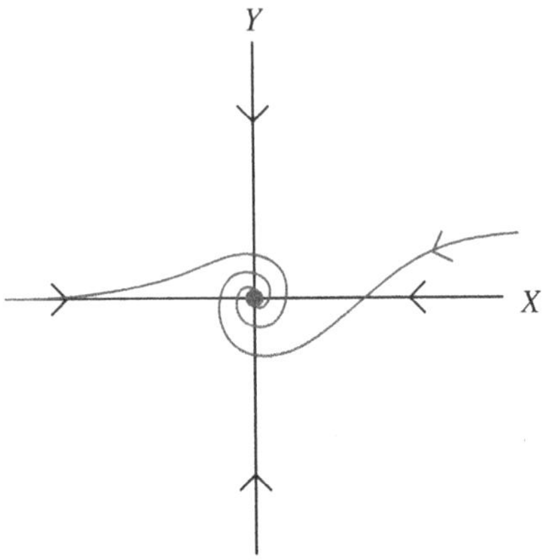

Fig. 1.4 Breakdown of a non-NHIM under a circular perturbation

but have smoothness C^r where r can be made arbitrarily large by decreasing the perturbation size. ○

1.2.2 Motivation for Noncompact NHIMs

Most of the literature on normal hyperbolicity and its applications treat compact NHIMs only. This excludes possibly interesting applications. Settings where a noncompact, general geometric version of normal hyperbolicity may be useful include chemical reaction dynamics [Uze+02] and problems in classical and celestial mechanics [DLS06].

We describe three examples where noncompactness naturally comes into play. The first example, a normally attracting cylinder, is set in Euclidean space. This example could be complicated a bit more by adding normal expanding directions to get a fully normally hyperbolic system. Such situations show up in Hamiltonian or reversible systems with invariant tori [Bro+09]. The second and third examples are set in ambient manifolds with nontrivial topology, thus motivating the need for a theory of noncompact NHIMs in such a geometric setting. The third example shows an application to classical mechanics and actually motivated this work.

Let us first treat a simple example.

Example 1.4 (*A normally attractive cylinder*) Let us consider the infinite cylinder $y^2 + z^2 = 1$ in $\mathbb{R}^3$. If we define a very simple dynamics by

$$(\dot{x}, \dot{r}, \dot{\theta}) = (0, r(1 - r), 1)$$

in cylindrical coordinates, then the cylinder is normally attractive and the motion on the cylinder consists of only periodic orbits, see Fig. 1.5.

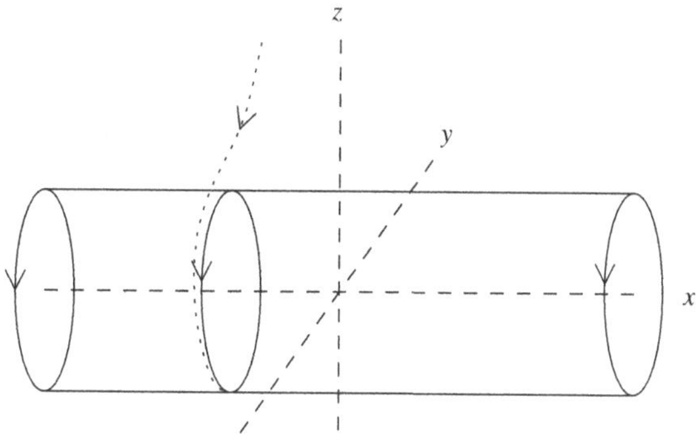

Fig. 1.5 A normally attracting cylinder

The dynamics on the cylinder is completely neutral, while it attracts in the normal direction with rate -1. Hence, there exists a unique persistent manifold diffeomorphic and close to the original cylinder. For any $k \geq 1$, the persistent manifold has C^k smoothness if the perturbation is chosen sufficiently small. The perturbed manifold must be uniformly close to the original cylinder; this rules out Example 3.9 of a cylinder with exponentially shrinking radius.

The dynamics on the persistent manifold can be perturbed in arbitrary ways. It could slowly spiral towards $x-$ infinity, or develop attracting and repelling periodic orbits on the cylinder. If the cylinder were higher dimensional, it could even become chaotic. ○

The previous example is set in Euclidean space, even though the invariant manifold has not completely trivial topology. Let us next consider a case where the ambient space is a manifold of nontrivial topology.

Example 1.5 (A pendulum with time-dependent perturbation) Consider a classical pendulum described by its angle $\theta \in S^1$ and angular velocity ω. The unstable top position $\theta = \pi$ is a hyperbolic fixed point, hence a special case of a NHIM.

The phase space $\mathrm{T}S^1$ is noncompact and nontrivial. Without perturbations we could restrict ourselves to a compact energy surface; if we add a general time-dependent perturbation, however, then energy is not preserved anymore and the full phase space must be considered. We should include the noncompact time interval $\mathbb{R}$ as well, leading to the full phase space $\mathrm{T}S^1 \times \mathbb{R}$. Then the unperturbed unstable top position corresponds to the one-dimensional NHIM

$$M = \{(\theta, \omega, t) \in \mathrm{T}S^1 \times \mathbb{R} \mid \theta = \pi,\ \omega = 0\}. \tag{1.3}$$

If the perturbation is globally sufficiently small in C^1 norm, then M uniquely persists into a manifold $\tilde{M}$ nearby. This means that under any small perturbations there exists a unique orbit of the pendulum that balances closely around the unstable top position for all time $t \in \mathbb{R}$. Note that this orbit stays uniformly close to the top position and that the perturbation need not depend (quasi-)periodically on time. ○

Remark 1.6 The nontrivial topology of S^1 can easily be undone by modeling the system on its universal cover $\mathbb{R}$. The example can be changed to a spherical pendulum modeled on S^2; this indicates more clearly the usefulness of having the theory available in spaces with nontrivial topology, although in this case we still only need to study the problem in a small neighborhood of the top position. It was pointed out to me by Robert MacKay that TS^2 is still a trivializable tangent bundle, and that more complex, non-trivializable examples including higher dimensional spheres can be constructed in the setting of chemical reaction dynamics. ◇

The last example actually motivated this work.

Example 1.7 (*Nonholonomic systems as singular perturbation limit*) Let a classical mechanical system be given by a smooth Riemannian manifold (Q, g) as configuration space and a Lagrangian $L\colon \mathrm{T}Q \to \mathbb{R}$. The vector field v on $\mathrm{T}Q$ is determined by the Lagrange equations of motion, given in local coordinates by

$$[L]^i = \frac{\mathrm{d}}{\mathrm{d}t}\frac{\partial L}{\partial \dot{x}_i} - \frac{\partial L}{\partial x_i} = 0. \tag{1.4}$$

A nonholonomic constraint can be placed on such a system by specifying a distribution[7] $\mathcal{D} \subset \mathrm{T}Q$ and adding reaction forces to $[L]$ according to the Lagrange–d'Alembert principle, that is, we require that a solution curve γ satisfies

$$[L](\gamma)(t) \in \mathcal{D}^0 \quad \text{and} \quad \dot{\gamma}(t) \in \mathcal{D} \quad \text{for all } t \in \mathbb{R} \tag{1.5}$$

where $\mathcal{D}^0 \subset \mathrm{T}^*Q$ denotes the annihilator of $\mathcal{D}$. This means that we restrict the velocities—but not the positions—of the system and adapt the vector field such that it preserves $\mathcal{D}$. Such constraints are called 'nonholonomic' if the distribution $\mathcal{D}$ is not integrable. This means that some small positional changes can only be obtained through long orbits due to the constraints. The prototypical example is that parallel parking a car a small distance sideways requires repeated turning and moving forward and backward.

As a concrete example of a nonholonomic system, let us consider a ball rolling on a flat surface. The possible positions of the ball are specified by $Q = SO(3) \times \mathbb{R}^2$, i.e. orientation and position in the plane. If we enforce the constraint that the ball can only roll and not slip, then its linear velocity is determined by its angular velocity $\omega \in \mathfrak{so}(3)$, thus we have

$$\mathfrak{so}(3) \times SO(3) \times \mathbb{R}^2 \cong \mathcal{D} \subset \mathrm{T}\big(SO(3) \times \mathbb{R}^2\big).$$

The addition of the nonholonomic reaction forces specified by the Lagrange–d'Alembert principle can be argued for on physical grounds, and some experimental verification has been done by Lewis and Murray [LM95] to check its correctness

---

7 Here, a distribution is meant in the sense of differential geometry as a subbundle of the tangent bundle, not a generalized function (nor a probability distribution).

against the alternative vakonomic principle. Still, it would be nice to rigorously derive these forces from fundamental principles; this would complement [RU57, Tak80, KN90] which showed this for holonomic constraints. The nonholonomically constrained system can be obtained from the unconstrained system by adding friction forces, see [Kar81, Bre81, Koz92]. Heuristically, one could say that if a rolling ball feels a strong contact friction force, then if this force is taken to infinity, it suppresses all slipping. This can be viewed as a singular perturbation limit, where $\mathcal{D}$ precisely is the invariant manifold, and it is normally attracting due to the dissipative friction force.

The cited works prove this result, but only asymptotically on finite time intervals. The extension of the theory of NHIMs to noncompact manifolds as developed in this work can be applied here. It allows one to improve upon this result and make it exact on infinite time intervals and general noncompact configuration spaces Q, as long as these satisfy the 'bounded geometry' condition. One could think, for example, of a gently sloping surface and a ball that is not perfectly round, or even a time-dependent perturbation, as long as it is uniformly bounded in time. ○

1.3 Historical Overview

As already mentioned, the theory of normally hyperbolic invariant manifolds is a generalization of the theory of hyperbolic fixed points. The study of these dates back to the beginning of the 20th century, or even the end of the 19th century. From 1892 onwards, Poincaré published his works "Les méthodes nouvelles de la mécanique céleste" [Poi92], in which he founded the theory of dynamical systems and famously studied the three-body problem. This triggered further research in nonlinear dynamical systems and persistence questions. Another important work published in the same year is "The general problem of the stability of motion" by Lyapunov; the original is in Russian, but translations in French [Lya07] and English [Lya92] are available. In this work, he introduced the concept of characteristic numbers, nowadays called 'Lyapunov exponents', to study 'conditional stability' of nonlinear differential equations at a fixed point. Conditional stability corresponds to the existence of stable (and unstable) linearized directions and Lyapunov proves the existence of a stable manifold by means of a series expansion under the assumption that the system is analytic.

In the beginning of the 20th century, the problem of stable manifolds was studied, without assuming analyticity, by Hadamard [Had01] and Cotton [Cot11]. Both Frenchmen applied different methods to obtain the stable and unstable manifolds of a hyperbolic fixed point. Later, the German mathematician Perron extended the ideas of Cotton to allow for generic complex eigenvalues, possibly of higher multiplicity, as long as the real parts of the eigenvalues are separated by zero (or even a number $r \neq 0$), see [Per29; Per30]. Hadamard's method is now named after him, and also known as the 'graph transform'. The other method was first formulated by Cotton, although the idea of exponential growth of solution curves can be traced to Lyapunov.

This method is commonly referred to as the Perron or Lyapunov–Perron method in the literature. This seems to pay too little credit to Cotton, even though Perron himself [Per29] does attribute the method to Cotton.[8]

From around 1960, renewed activity in the area of hyperbolic dynamics led to the generalization of the theory of (un)stable manifolds for hyperbolic fixed points to persistence and (un)stable fibrations for normally hyperbolic invariant manifolds. Many authors have contributed to this subject, culminating in the seventies in the works by Fenichel [Fen72] and Hirsch, Pugh, and Shub [HPS77]. These two works formulate the theory slightly differently, but in broad generality and can be viewed as the basic references nowadays; references to earlier works can be found in both. Both Fenichel and Hirsch, Pugh, and Shub use Hadamard's graph transform as their fundamental tool. In these works, compactness of the invariant manifold is a basic assumption. Noncompact, immersed manifolds are considered in Ref. [HPS77, Sect. 6], albeit under the assumption that the immersion image is compact again.

The theory of normal hyperbolicity has seen some interesting developments since these foundational works, and the applications have slowly started to flourish, see [Wig94] for a list of subjects. A major development was the generalization to semi-flows in Banach spaces. This situation can arise when one wants to study partial differential equations as ordinary differential equations on appropriate function spaces. This technique has been applied to PDEs such as the Navier–Stokes or reaction-diffusion equations.

In his book on parabolic PDEs, Henry extended the Perron method to apply to semi-flows with a NHIM given as the horizontal submanifold[9] $X \times \{0\}$ in a product $X \times Y$ of Banach spaces [Hen81, Chap. 9]. Henry's idea is to linearize only the normal directions, but keep the horizontal flow along M in its general, nonlinear form, while at the same time splitting the Perron contraction map into a two-stage contraction map on horizontal and vertical curves separately. Henry obtains $C^{1,\alpha}$ smoothness only. In the series of papers [BLZ98; BLZ99; BLZ08], Bates, Lu, and Zeng study more general NHIMs of semi-flows in Banach spaces. They employ Hadamard's graph transform and allow so-called 'overflowing invariant manifolds', as in [Fen72]. They also allow the NHIM to be noncompact and an immersed instead of an embedded submanifold. In [BLZ99] the unperturbed NHIM is assumed to be C^2 to obtain C^1 persistence results, for the technical reason of constructing C^1 normal bundle coordinates. In their later paper [BLZ08], this technicality is overcome,[10] and existence of a NHIM is even proven when sufficiently close, approximately normally hyperbolic invariant manifolds exist; the persistence result is then obtained for compact NHIMs only, to circumvent the C^2 assumption.

Vanderbauwhede and Van Gils [VG87; Van89] introduced the technique of considering a scale (family) of Banach spaces of curves with exponential growth, and

---

[8] These facts were pointed out to me by Duistermaat.

[9] Henry actually has reversed notation where the 'vertical' manifold $Y \times \{0\}$ is the NHIM.

[10] Their Hypothesis (H2) that a certain approximate splitting like (1.9) "does not twist too much", can be obtained from uniform Lipschitz continuity of the tangent spaces of the invariant manifold. I am not sure if this is a significantly weaker hypothesis. See also the discussion in Remark 3.13.

using the fiber contraction theorem (see Appendix D), proved smoothness of center manifolds with the Perron method. Although not the same, center manifolds have many properties in common with NHIMs and Sakamoto [Sak90] has built upon the works of Henry and Vanderbauwhede and Van Gils to prove persistence and C^{k-1} smoothness for singularly perturbed systems in a finite-dimensional $\mathbb{R}^m \times \mathbb{R}^n$ product space setting. The loss of one degree of smoothness is again due to the construction of normal bundle coordinates, although this fact is obscured by the explicit $\mathbb{R}^m \times \mathbb{R}^n$ setting.

Singularly perturbed, or, slow-fast systems are another important class of applications. These describe systems where the dynamics is governed by multiple, separate time scales, or when a system can be viewed as an approximation of an idealized, restricted system. Singularly perturbed systems can be studied using the theory of normal hyperbolicity by turning them into a regular perturbation problem via a rescaling of time, see foundational work by Fenichel [Fen79] or the more introductory expositions [Jon95; Kap99; Ver05].

1.4 Comparison of Methods

There are two well-known methods for proving the existence and smoothness of invariant manifolds in hyperbolic-type dynamical systems. The Hadamard graph transform and the variation of constants method, also known as the (Lyapunov–) Perron method. Variations of both have been applied in many situations with some form of hyperbolic dynamics. This ranges from the relatively simple problem of finding the stable and unstable manifolds of a hyperbolic fixed point, to center manifolds, partially hyperbolic systems, and normally hyperbolic systems. The quote of Anosov [Ano69, p. 23] that "every five years or so, if not more often, someone 'discovers' the theorem of Hadamard and Perron, proving it either by Hadamard's method of proof or by Perron's" is nowadays probably familiar to many researchers in these areas; it illustrates the pervasiveness of these methods.

In this section, I describe the ideas that are common to both methods, as well as their differences. I hope to elucidate the merits and weak points of both methods, especially when applied to normally hyperbolic systems. Basically they seem to be able to produce the same conclusions, but each method takes a different viewpoint to the problem.

Let us first identify some basic common ideas. As a sample problem, we consider finding the invariant unstable manifold W^U of a hyperbolic fixed point, positioned at the origin of $\mathbb{R}^n$. The system is defined by either a diffeomorphism Φ in the discrete case, or a flow Φ^t in the continuous case. Both methods use the splitting of the tangent space into stable and unstable directions:

$$\mathrm{T}_0\mathbb{R}^n \cong \mathbb{R}^n = U \oplus S.$$

Let (x^+, x^-) denote coordinates in $U \oplus S$ according to projections π^+, π^- from $\mathbb{R}^n$ onto the unstable and stable directions U and S, respectively. We shall use the notation $\Phi_\pm = \pi^\pm \circ \Phi$.

1.4.1 Hadamard's Graph Transform

The graph transform is due to Hadamard. His paper [Had01] (in French, 4 pages) can be used as a concise and basic introduction to the graph transform, applied to the stable and unstable manifolds of a hyperbolic fixed point. He does not prove smoothness or even continuity of these invariant manifolds, although continuity could easily be concluded by introducing the Banach space of bounded continuous functions with supremum norm.

The basic idea of the graph transform is to view the unstable manifold W^U as the graph of a function $g : U \to S$. The graph, as a set, is invariant under Φ (or e.g. Φ^1 in the continuous case). The diffeomorphism Φ can also be interpreted as a map acting on functions g through its action on their graphs. This induces a mapping

$$T : g \mapsto \tilde{g} \qquad \text{implicitly defined by} \qquad \tilde{g}\big(\Phi_+(x, g(x))\big) = \Phi_-(x, g(x)). \tag{1.6}$$

Thus, by definition, any point $(x, g(x))$ on the graph of g gets mapped to a point $(x', \tilde{g}(x'))$ on $\mathrm{Graph}(\tilde{g})$. The map T turns out to be well-defined and a contraction on functions $U \to S$ that are sufficiently small in Lipschitz norm. The graph of the unique fixed point $g^\star$ of T must correspond to the unstable manifold, that is, $W^U = \mathrm{Graph}(g^\star)$.

By considering the invariant sets, this method focuses on the geometry of the problem. The method uses a diffeomorphism map Φ; the continuous case can be studied by considering the flow map Φ^t for a fixed time t. The diffeomorphism can easily be studied locally in charts on a manifold. Therefore this method lends itself well to the generalized setting of normally hyperbolic invariant manifolds, where the invariant manifold is intrinsically a global object. Even if this global object is nontrivial, it can still be studied in local charts.

1.4.2 Perron's Variation of Constants Method

This method is commonly referred to as the Perron or Lyapunov–Perron method. Although in the literature this is attributed to Perron [Per29], he in turn cites Cotton [Cot11] for the main idea.

This method focuses on the behavior of solution curves. The solutions on the unstable manifold are precisely characterized by the fact that they stay bounded under backward evolution. In the following, we explain the Perron method for the continuous case.[11] We adopt the notation from the graph transform setting. A contraction operator T is constructed via a variation of constants integral. The nonlinear part of the vector field is viewed as a perturbation of the linear part. The integral equation is split into the components along the stable and unstable directions. Then the integration of the unstable component is switched from the interval $[0, t]$ to $[-\infty, t]$, and only bounded functions are considered. Writing the vector field $v(x) = \mathrm{D}v(0) \cdot x + f(x)$ in linearized form with nonlinearity f, this leads to the following contraction operator on curves $x = (x^+, x^-) \in C^0([-\infty, 0]; \mathbb{R}^n)$:

$$T\colon \bigl(x^+(t), x^-(t)\bigr) \mapsto \Bigl(x_0^+ - \int_t^0 \mathrm{D}\Phi_+^{t-\tau}(0)\; f_+\bigl(x^-(\tau), x^+(\tau)\bigr)\,\mathrm{d}\tau\;, \\ \int_{-\infty}^t \mathrm{D}\Phi_-^{t-\tau}(0)\; f_-\bigl(x^-(\tau), x^+(\tau)\bigr)\,\mathrm{d}\tau\Bigr). \tag{1.7}$$

This mapping T is well-defined and a contraction on curves $x \in C^0([-\infty, 0]; \mathbb{R}^n)$ whose stable component x^- is bounded and sufficiently small. Note that T does not depend on the stable component x_0^- of the initial conditions anymore. The fixed point of T is a solution curve on W^U with x_0^+ given as a parameter. The unstable manifold is described, finally, by evaluating the stable component at zero, leading to a graph

$$g\colon U \to S\colon x_0^+ \mapsto x^-(0).$$

First of all, it must be noted that this method requires f to be small in C^1-norm. We can make f small by restricting to a sufficiently small neighborhood of the origin and cutting off f outside of it. This cut-off does not influence the results: due to the boundedness condition, curves x stay in the neighborhood. The method can be generalized to a separation of stable and unstable spectra (i.e. a dichotomy) away from the imaginary axis,[12] and for example be applied to show existence of center manifolds. In that case, uniqueness is lost as solutions will generally run out of small neighborhoods. This makes the Perron method not directly applicable to normally hyperbolic invariant manifolds. The center direction corresponds to the invariant manifold, but solution curves are global objects that cannot be treated locally.

The Perron method can be extended to overcome this problem. Henry [Hen81, Chap. 9] linearizes the vector field only in the normal directions of the invariant manifold. Henry uses a two-step contraction scheme, but this can be reduced to a single contraction $T = T_- \circ T_+$ that is a composition of two maps. The maps $T_\pm$ are

---

[11] Contrary to the graph transform (which is only intrinsically defined for mappings), the Perron method can be formulated both for flows and discrete mappings. For the discrete case, the integral must be replaced by a sum, the mapping Φ must be split into a linear and nonlinear part, and the linearized flow must be replaced by iterates of the linearized mapping. See for example [APS02; PS04].

[12] This is for the continuous case. The imaginary axis of the spectrum of a vector field corresponds (via the exponential map) to the unit circle for the spectrum of a diffeomorphism in the discrete case.

essentially the components of (1.7). Still, the results obtained are not quite as general as those obtained with the graph transform. For the graph transform, the condition of normal hyperbolicity can be formulated in terms of the ratio of the normal and tangential growth rates of the flow along orbits, while for the Perron method it must be formulated in terms of the ratio of global growth rates. This less general assumption is required because the contraction operator (1.7) is studied on spaces of solution curves with a fixed exponential growth behavior, see Definition 1.17.

Explicit time dependence can be added to the Perron method with only trivial modifications. This allows one to study hyperbolic fixed points in non-autonomous systems.[13] An application is the study of invariant fibrations of, for example, normally hyperbolic invariant manifolds. These have fibered stable and unstable manifolds. Points in a single fiber are characterized by the unique orbit on the normally hyperbolic invariant manifold they are exponentially attracted to under forward or backward evolution, respectively. Finding these fibers is turned into a non-autonomous hyperbolic fixed point problem by following a point on the invariant manifold.

1.4.3 Smoothness

In the truly hyperbolic case—when the stable and unstable spectra are separated by a neighborhood of the imaginary axis—the Perron method allows for a direct proof of smoothness of the manifolds W^U and W^S, see [Irw70; Irw72] where this is formulated for discrete systems. One first verifies that the contraction operator T is as smooth as the system, still acting on continuous curves x. Then, by an implicit function theorem argument, the fixed point depends smoothly on the (partial) initial value parameter x_0^+. To the best of my knowledge, there is no similarly simple approach for the graph transform. The contraction map acts directly on graphs g, so to obtain smoothness, one must consider the maps $g \in C^k(U; S)$. A direct estimate of contractivity in C^k-norm requires higher than k-th order Lipschitz estimates on the system.

When the spectra are not separated by the imaginary axis—this occurs for example in normally hyperbolic systems—things become more complicated. The spectral gap condition defines an intrinsic upper bound for the smoothness that one can generically expect for a system, as was seen in Example 1.1. Both methods apply induction over the smoothness degree in their proof. Formal derivatives of the contraction map T are constructed. These are again contractions, but now on higher derivatives of the fixed point mapping, while fixing the derivatives below. Finally, the fiber contraction theorem (see Appendix D) can be used to conclude that these higher order derivatives converge to a fixed point, jointly with all lower orders.

[13] The term 'fixed point' in the context of a non-autonomous system is not definable in a coordinate-free way: any orbit of the system can be made into a fixed point under a suitable time-dependent coordinate transformation. However, there may be a preferred "time-independent" coordinate system. Moreover, the hyperbolicity of an orbit with respect an intrinsic metric is independent of a choice of coordinates.

Explicit calculation of higher derivatives of T is very tedious; one should focus on their form as dictated by Proposition C.3. For the graph transform, the relevant terms that one obtains from (1.6) are, ignoring arguments,

$$\mathrm{D}^k\tilde{g} \cdot \big(\mathrm{D}_1\Phi_+ + \mathrm{D}_2\Phi_+\,\mathrm{D}g\big)^k + \cdots = \mathrm{D}_2\Phi_- \cdot \mathrm{D}^k g + \cdots$$

This leads to a contraction when $\|\mathrm{D}_2\Phi_-\| \cdot \left\|\mathrm{D}_1\Phi_+^{-1}\right\|^k < 1$. The limit on k precisely corresponds to the spectral gap condition, at least when we replace Φ by a sufficiently high iterate Φ^N of itself, or in the continuous case, if we take the flow map Φ^t at a sufficiently large time t.

For the Perron method, the essential form of the derivatives of T is

$$\mathrm{D}^k T(x)\big(\delta x_1, \ldots, \delta x_k\big)(t) = \int \mathrm{D}\Phi^{t-\tau}(0) \cdot \mathrm{D}^k f(x(\tau))\big(\delta x_1(\tau), \ldots, \delta x_k(\tau)\big)\,\mathrm{d}\tau. \tag{1.8}$$

The solution curve x as well as its variations δx_i are of growth order $e^{\rho t}$, so the variation of f in the integrand is of growth order $e^{k\rho t}$, even if $\mathrm{D}^k f$ itself is bounded. This means that k-th order variations must be considered in spaces of growth order $e^{k\rho t}$ and $\mathrm{D}^k T$ is only contractive on such spaces if both ρ and $k\,\rho$ are contained in the spectral gap.

1.5 Bounded Geometry

The main results of this work are formulated in a geometric context on differentiable manifolds. Already in [Fen72; HPS77] the results are formulated in such a context. This allows for more general situations than choosing $\mathbb{R}^n$ as ambient space. In the compact case, it does not require a change in the basic proofs (as can be seen from the approach taken in [Fen72]), but it does bring in some additional formalism. On the other hand, pre-existing work on noncompact NHIMs [Hen81; Sak90; BLZ08; BLZ99] explicitly assumes that the ambient space is Euclidean or Banach.

It turns out that if one switches to a noncompact setting in manifolds, then some fundamental new ingredients must be added. Already in Euclidean space one must assume uniformity of the vector fields as a replacement for compactness. Similar additional uniformity assumptions are required for the ambient space—these do not manifest themselves in Euclidean space. First, a choice of Riemannian metric (or possibly a weaker form: a Finsler structure) is required since not all metrics are equivalent anymore on a noncompact manifold, see Example 3.6. As an extension, Example 3.7 shows that one cannot reduce the noncompact to a compact case by compactification. Secondly, the ambient manifold and functions on it should satisfy uniformity criteria that can be captured in terms of 'bounded geometry'.[14] For full

[14] We do not claim that bounded geometry is a necessary condition to generalize the theory of normal hyperbolicity to noncompact ambient spaces, only that it is sufficient. Section 3.3 does contain some examples, though, that indicate that some form of bounded geometry is necessary.

details see Sect. 3.3 on compactness and uniformity and Chap. 2 on bounded geometry. Let us just give a quick overview here.

A Riemannian manifold has bounded geometry, loosely speaking, if it is globally, uniformly well-behaved. More precisely, its curvature must be bounded and the injectivity radius must be bounded away from zero, see Definition 2.1. Then there exists a preferred set of so-called normal coordinate charts for which coordinate transition maps are uniformly continuous and bounded, smooth functions. That is, in k-th order bounded geometry we have a C^k uniform atlas. As a consequence, uniformly continuous and bounded submanifolds, vector fields, and other objects can be defined and manipulated in a natural way in terms of these coordinates. Note that $\mathbb{R}^n$ and compact manifolds have bounded geometry, see Example 2.3. Together with corollaries 3.4 and 3.5 of the main theorem, this shows that bounded geometry provides a natural generalization to the known settings of compact and Euclidean spaces.

We use bounded geometry to obtain boundedness estimates on holonomy, see Sect. 2.2. This is a fundamental ingredient in our proof of smoothness of the perturbed manifold. Finally, we present more technical results in bounded geometry: a uniform tubular neighborhood, uniform smoothing of submanifolds, and a trivializing embedding of the normal bundle. We use these to reduce the full problem of persistence of a normally hyperbolic submanifold M in an ambient manifold Q to the trivialized situation $X \times Y$, where M is represented by the graph of a small function $h\colon X \to Y$ and Y is a vector space. Uniformity permeates all these constructions in order to obtain uniform estimates required for the persistence proof in the trivialized setting.

1.6 Problem Statement and Results

The main problem in this work is the persistence of normally hyperbolic invariant manifolds under small perturbations of the dynamical system. That is, given a flow Φ^t defined by some vector field v and a normally hyperbolic invariant submanifold M, we want to show that for any vector field $\tilde{v}$ sufficiently close to v, there exists a unique manifold $\tilde{M}$ close to M that is invariant under the flow of $\tilde{v}$; moreover we would like to show that $\tilde{M}$ is normally hyperbolic again. To make this statement precise, we need to define a lot of things: first of all, we need to rigorously define normal hyperbolicity. Secondly, the statements about vector fields and manifolds being 'close' need to be formalized and finally, we need to specify the ambient space Q on which the system is defined.

We start with a Riemannian manifold (Q, g) as ambient space and a submanifold M. For technical reasons this manifold is assumed to be complete and of bounded geometry (or at least in a $\delta > 0$ neighborhood of M, since the whole analysis can be restricted to such a neighborhood). Basically, these conditions impose uniformity of the space, and fit in the principle of replacing compactness by uniform estimates, see Sect. 1.3 and Chap. 2 for more details. Note that $Q = \mathbb{R}^n$ with the standard Euclidean metric is an easy (and typical) special case.

Let $v \in \mathfrak{X}(Q)$ be a vector field on Q with $v \in C^{k,\alpha}_{b,u}$, that is, v up to its k-th derivative is uniformly continuous and bounded, and α-Hölder continuous if $\alpha \neq 0$. On $\mathbb{R}^n$ these statements make immediate sense; on general manifolds Q, results from Chap. 2 are required, in particular Definition 2.9, to make sense of uniform boundedness and continuity by means of normal coordinates. Let $\tilde{v}$ be another such vector field. The closeness of v and $\tilde{v}$ will be measured using supremum norms. The C^1-norm is required to be small for the persistence result. Thus, even though we consider the space of $C^{k,\alpha}$ bounded vector fields, we endow this space with a C^1 topology. See Sect. 1.7 for some more remarks on this topology and a comparison with standard topologies on noncompact function spaces. If we assume that $\tilde{v} - v$ is small in $C^{k,\alpha}$-norm as well, then $\tilde{M}$ will be $C^{k,\alpha}$-close[15] to M. These C^1 and C^k norm requirements and results are direct analogues of those in the implicit function theorem.

Finally, we define normal hyperbolicity of a submanifold M with respect to a continuous dynamical system $(\mathbb{R}, Q, \Phi)$. The flow Φ^t should have a domain of definition containing at least a neighborhood of the invariant manifold M. This definition is easily adapted to the discrete case of a diffeomorphism $\Phi \colon Q \to Q$; simply replace $t \in \mathbb{R}$ by $t \in \mathbb{Z}$ as iterated powers of Φ.

Definition 1.8 **(Normally hyperbolic invariant manifold)** *Let (Q, g) be a smooth Riemannian manifold, $\Phi^t \in C^{r \geq 1}$ a flow on Q, and let $M \in C^{r \geq 1}$ be a submanifold of Q. Then M is called a normally hyperbolic invariant manifold of the system (Q, Φ^t) if all of the following conditions hold true:*

1. *M is invariant, i.e. $\forall\, t \in \mathbb{R} \colon \Phi^t(M) = M$;*
2. *there exists a continuous splitting*

$$\mathrm{T}_M Q = \mathrm{T}M \oplus E^+ \oplus E^- \tag{1.9}$$

of the tangent bundle $\mathrm{T}Q$ over M with globally bounded, continuous projections π_M, π_+, π_- and this splitting is invariant under the tangent flow $\mathrm{D}\Phi^t = \mathrm{D}\Phi^t_M \oplus \mathrm{D}\Phi^t_+ \oplus \mathrm{D}\Phi^t_-$;

3. *there exist real numbers $\rho_- < -\rho_M \leq 0 \leq \rho_M < \rho_+$ and $C_M, C_+, C_- > 0$ such that the following exponential growth conditions hold on the various subbundles:*

$$\begin{aligned}
&\forall\, t \in \mathbb{R},\ (m, x) \in \mathrm{T}M\colon && \left\| \mathrm{D}\Phi^t_M(m)\, x \right\| \leq C_M\, e^{\rho_M |t|}\, \|x\|,\\
&\forall\, t \leq 0,\ (m, x) \in E^+\colon && \left\| \mathrm{D}\Phi^t_+(m)\, x \right\| \leq C_+\, e^{\rho_+ t}\, \|x\|,\\
&\forall\, t \geq 0,\ (m, x) \in E^-\colon && \left\| \mathrm{D}\Phi^t_-(m)\, x \right\| \leq C_-\, e^{\rho_- t}\, \|x\|.
\end{aligned} \tag{1.10}$$

---

[15] We actually only obtain C^k closeness for integer $k \leq r - 1$ where r is the ratio in the spectral gap condition (1.11). This is probably an artifact of the techniques we used, while $C^{k,\alpha}$ closeness with $k + \alpha = r$ should be obtainable.

These exponential estimates imply that the tangent flow $D\Phi^t$ must contract at a rate of at least ρ_- along the stable complementary bundle E^-, expand[16] as $e^{\rho_+ t}$ along the unstable bundle E^+, and may not expand or contract at a rate faster than $\pm\,\rho_M$, respectively, tangent along TM.

Remark 1.9 We added the condition that the projections π_M, π_+, π_- are globally bounded. This is a natural extension to the noncompact case, and is automatically satisfied in case M is compact.

Remark 1.10 This definition of normal hyperbolicity is not as general as could be. Fenichel [Fen72, pp. 200–204] defines normal hyperbolicity in terms of 'generalized Lyapunov type numbers'. It follows from his uniformity lemma that these are essentially exponentiated versions of our Lyapunov exponents ρ. For example, his ν is equivalent to our $e^{-\rho_+}$. But Fenichel defines σ in terms of the ratio ρ_M/ρ_+ along orbits in M. His definition allows the expansion rate along TM to be large, for example, as long as the expansion rate along E^+ is large enough to keep the ratio $\sigma(m)$ bounded, *along the orbit through m*. The definitions in [HPS77, Mañ78, BLZ08] are equivalent in the compact context to the one in [Fen72]. Mañé's work shows that this definition is as general as possible, see below. ◇

When M is compact, normal hyperbolicity is a sufficient condition for the existence of a persistent manifold $\tilde{M}$ for a system generated by $\tilde{v}$ if $\|\tilde{v} - v\|_1$ is sufficiently small. Conversely, Mañé [Mañ78] has proved that normal hyperbolicity (in the sense of e.g. Fenichel's definition) is also necessary: if a compact invariant manifold M is persistent under any C^1 small perturbation, then M is normally hyperbolic (see also Example 1.2 and the clear exposition in the introduction of [Fen72]). Definition 1.8, however, only guarantees C^1 smoothness for the perturbed manifold $\tilde{M}$. To obtain higher order smoothness, a more stringent condition of r-normal hyperbolicity must be satisfied.

Definition 1.11 (**r-normally hyperbolic invariant manifold**) *A manifold M is called r-normally hyperbolic with $r \geq 1$ a real number, if it satisfies $M \in C^r$ and the conditions in Definition 1.8, but with the stronger inequalities*

$$\rho_- < -r\,\rho_M \leq 0 \leq r\,\rho_M < \rho_+. \tag{1.11}$$

This means that the normal expansion and contraction must not just dominate the tangential ones, but do so by a factor r. For $r = 1$ we recover the original definition, while the generalized inequality (1.11) is called the spectral gap condition. If M is r-normally hyperbolic and v and the perturbation $\tilde{v}$ are C^r as well, then the persistent manifold $\tilde{M}$ is C^r smooth again. The example in Sect. 1.2.1 shows that this spectral gap condition is sharp: even when everything is C^∞, the perturbed manifold $\tilde{M}$ in

[16] Note that expansion along E^+ could also be formulated as $\|D\Phi^t(m)\,x\| \geq C_+\,e^{\rho_+ t}\,\|x\|$ for $t \geq 0$ and $(m, x) \in E^+$. This is equivalent to the condition as stated, which says that there is contraction for $t \leq 0$, that is, in backward time. This latter formulation is preferable because it is the form required in estimates.

that example is only C^r when no more than r-normal hyperbolicity holds. Note that r can be interpreted as a 'fractional differentiability degree' when writing $r = k + \alpha$ with integer $k \geq 1$ the normal degree of differentiability and $0 \leq \alpha \leq 1$ an additional Hölder continuity exponent.

Remark 1.12 We explicitly exclude the case $r = \infty$ from Definition 1.11, even though the spectral gap condition (1.11) could hold for $r = \infty$, if $\rho_M = 0$. The reason is that one can generally not expect to obtain a persistent manifold $\tilde{M} \in C^\infty$ in this case. Even though for any order $r < \infty$ there exist persistent manifolds $\tilde{M} \in C^r$ for sufficiently small perturbations, the maximum perturbation size generally depends on r and may shrink to zero when $r \to \infty$. See Example 1.3 and the example in [Str79] for the closely related case of center manifolds.

On the other hand, it is shown in [HPS77] that there is forced smoothness. If $M \in C^1$ is an r-NHIM and the system is C^r, then M must be C^r. This also holds in our noncompact setting, see Remark 3.3, 10 for a sketch of its proof. ◇

With these preliminary definitions in place, we are now ready state our main theorem; it is restated in Chap. 3. We should point out that M is not required to be an embedded submanifold; immersions are allowed as well, see Sect. 1.6.2. For the details of the smoothness notation $C^{k,\alpha}_{b,u}$ on manifolds we refer to definitions 2.9 and 2.21.

Theorem 1.1 (Persistence of noncompact NHIMs in bounded geometry) *Let $k \geq 2$, $\alpha \in [0, 1]$ and $r = k + \alpha$. Let (Q, g) be a smooth Riemannian manifold of bounded geometry and $v \in C^{k,\alpha}_{b,u}$ a vector field on Q. Let $M \in C^{k,\alpha}_{b,u}$ be a connected, complete submanifold of Q that is r-normally hyperbolic for the flow defined by v, with empty unstable bundle, i.e.* $\operatorname{rank}(E^+) = 0$.

Then for each sufficiently small $\eta > 0$ there exists a $\delta > 0$ such that for any vector field $\tilde{v} \in C^{k,\alpha}_{b,u}$ with $\|\tilde{v} - v\|_1 < \delta$, there is a unique submanifold $\tilde{M}$ in the η-neighborhood of M, such that $\tilde{M}$ is diffeomorphic to M and invariant under the flow defined by $\tilde{v}$. Moreover, $\tilde{M}$ is $C^{k,\alpha}_{b,u}$ and the distance between $\tilde{M}$ and M can be made arbitrarily small in C^{k-1}-norm by choosing $\|\tilde{v} - v\|_{k-1}$ sufficiently small.

This result generalizes the well-known results in [Fen72, HPS77] to the case of noncompact submanifolds of Riemannian manifolds. Again, our definition of normal hyperbolicity is slightly less general than the definitions used in these works. We also assumed that only the stable bundle E^- is present, see also Sect. 4.4; note that we thus only have the spectral gap condition $\rho_- < -r\,\rho_M$ with $\rho_M \geq 0$. See also the restatement of this theorem on page 76 and the list of remarks 3.3 for more details.

We borrow the idea to generalize the Perron method to NHIMs from Henry [Hen81], and use the techniques of Vanderbauwhede and Van Gils [VG87] (see [Van89] for a clear presentation) for proving higher order smoothness. This is similar, but developed independently from Sakamoto's work [Sak90] in which he used the same ideas to study singular perturbation problems. We improve these results in a couple of ways. First of all, we simplify the basics of the proof by reducing the two-step contraction argument to a single contraction mapping, still written as a composition of two separate maps acting on horizontal curves in M and vertical curves in the normal bundle fiber, respectively. More importantly, we remove the restriction of a trivial product structure $X \times Y$. Thus, we neither require M to have a global chart in a Banach space X, so M need not be topologically trivial, nor do we require a global product, so the normal bundle of M need not be trivial either. On the other hand, the results by Bates, Lu, and Zeng also allow M to be a general submanifold, but still assume the ambient space to be a Banach space. Our results are for finite dimensional, but not necessarily linear, Riemannian ambient spaces. In their paper [BLZ08], they only require an approximate NHIM for finding a persistent invariant manifold. We use this idea as well (see the setup of h small in the formulation of Theorem 3.2), but we do not expand this idea any further. Finally, this work was initiated from the (unfortunately never published) preprint by Duistermaat on stable manifolds [Dui76].

It seems to be a well-known belief by many experts that the theory of normal hyperbolicity can be extended to a general noncompact setting [DLS06, p. 165]. The idea is to replace compactness by uniform estimates. An important conclusion to be drawn from the present work is that indeed this principle holds, but probably in a more strict way than one would naively realize. Uniform estimates are not only required for the vector field defining the system, but for the underlying ambient space as well, in terms of bounded geometry. This becomes clear only when one leaves the context of Euclidean ambient spaces, which trivially have bounded geometry. On a Riemannian manifold, already the very definition of uniform continuity of a vector field v and its derivatives requires some aspects of bounded geometry. It should be noted though, that we do not prove that bounded geometry is a strictly necessary condition for persistence of NHIMs; nonetheless, the results do suggest that persistence of NHIMs may break down in 'unbounded geometry', see Sect. 3.3.

In Sect. 3.2 we present an outline of the proof and how it is reduced to a more basic setting $M' \times Y$ of a trivial normal bundle. Here M' is a smoothed version of M to rectify an artificial loss of smoothness, as occurs e.g. in Ref. [Sak90]. Below we present some extensions to the main Theorem 1.1 above.

1.6.1 Non-Autonomous Systems

Our main theorem can be trivially extended to the non-autonomous, time-dependent case. First, extend the configuration space with time t as additional variable, i.e. $\hat{Q} = Q \times \mathbb{R}$, and add the equation $\dot{t} = 1$. If the original system was time-independent, then $\hat{M} = M \times \mathbb{R}$ is a NHIM for the extended system, and all uniform assumptions still hold, since the flow along the time direction is neutral and trivial. Note that this argument does not work in the classical theory as $\hat{M}$ is not compact.[17] Now we can

[17] If the perturbation is time-dependent, but in an (almost) periodic way, then this can still be treated in the compact setting. One can extend the configuration space with the circle S^1 (or an n-torus in the almost periodic case).

make any C^1 small perturbation, and obtain a persistent manifold $\tilde{M}$ in the extended configuration space. The perturbation is allowed to be generally time-dependent, as long as it is uniform in time, including derivatives. The resulting manifold $\tilde{M}$ will still be invariant and close to the original M, although it will depend on time. That is, if we assume local coordinates $(x, y) \in \mathbb{R}^n \times \mathbb{R}^m$ for Q such that $M = \mathbb{R}^n \times \{0\}$ locally, then we can write $\tilde{M} = \mathrm{Graph}(h)$ for a function

$$h\colon \mathbb{R}^n \times \mathbb{R} \to \mathbb{R}^m, \quad y = h(x, t).$$

In other words, $\tilde{M}$ can be viewed as a graph over M (i.e. a section of the normal bundle), but this graph now additionally depends on time. The manifold $\tilde{M}$ itself is again normally hyperbolic when viewed in the extended space $Q \times \mathbb{R}$, see also Sect. 4.1.

Such time-dependent invariant manifolds are called 'integral manifolds'. These have been studied as non-autonomous generalizations of stable and unstable manifolds of hyperbolic fixed points [Pal75], but also as generalizations of compact NHIMs [Hal61, Yi93]. The theory of noncompact NHIMs allows one to treat all such integral manifolds in the same way as the autonomous case. One can, for example, also start with an integral manifold that is normally hyperbolic: it will persist just as well.

1.6.2 Immersed Submanifolds

In the main Theorem 1.1, we intentionally do not precisely state in what sense M is a submanifold of Q. The implicit assumption that M is an embedded submanifold can be weakened to M being an immersion, see also [HPS77, Sect. 6] and [BLZ99]. That is, M can be viewed as an abstract manifold together with an immersion map $\iota\colon M \to Q$ that need not be injective. This does not affect the theory as long as ι is still locally injective: $\iota(M)$ including a neighborhood modeled on its normal bundle N can be pulled back via the immersion ι to the abstract M. All local properties are preserved, so we can study the system via this 'covering'. We may not always make a clear distinction between the abstract manifold M and its immersed image $\iota(M) \subset Q$; the discussion below shows that this distinction is not really necessary, as long as we do not consider perturbations.

For a generic immersion one could expect a picture as in Fig. 1.6, where the immersed manifold intersects itself transversely. Such situations cannot occur if M is a NHIM. This follows from the exponential growth rates along tangent and normal bundles of M. Let $m \in \iota(M)$ be an intersection point of two preimages $m_1, m_2 \in M$. If the tangent spaces along M at m_1 and m_2 are embedded differently into $\mathrm{T}_m Q$, then one could find $x \in \mathrm{Im}\big(\mathrm{D}\iota(m_1)\big) \setminus \mathrm{Im}\big(\mathrm{D}\iota(m_2)\big)$. This would imply that x has a component in N_{m_2} and give contradictory growth rates for $\mathrm{D}\Phi^t(m) \cdot x$ depending on whether we view m as image of m_1 or m_2, as the orbit of $m \in \iota(M)$ is uniquely defined. Hence, at each point $m \in \iota(M)$ the tangent spaces $\mathrm{D}\iota(m_i)$ of all preimages

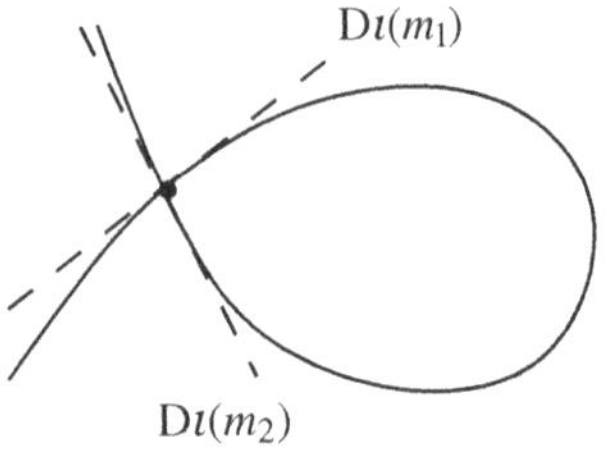

Fig. 1.6 An immersion with a transverse intersection

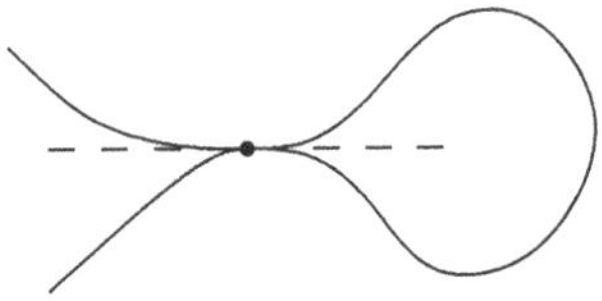

Fig. 1.7 An allowed immersion with tangential intersection

$m_i \in \iota^{-1}(m)$ must coincide, see Fig. 1.7. Stated more abstractly, M must have contact of order one with itself. More generally it holds that an immersed k-NHIM has contact of order k with itself,[18] see [HPS77, p. 68].

Next, each maximal set of $\iota(M)$ with constant number of preimages[19] $p \in \mathbb{N} \cup \{\infty\}$,

$$M_p = \{m \in \iota(M) | \# \iota^{-1}(m) = p\}, \tag{1.12}$$

is an invariant subset of $\iota(M)$. This is again due to uniqueness of the flow. If an orbit would cross into a set of different preimage number, then a least one of the 'lifts' of this orbit from $\iota(M)$ to the 'cover' M would have to enter or leave M. This cannot happen as M itself is invariant. Hence, the conclusion is that self-intersections of $\iota(M)$ must be invariant.

Immersed NHIMs may occur on themselves, or appear as a persistent manifold under perturbation from an embedded manifold. An example of an embedded noncompact NHIM that collapses under a small perturbation into an immersed manifold can be found in Sect. 3.3. The same can happen with an immersed manifold with compact image. The following example is taken from [HPS77, p. 130] and shows that the injection map is relevant for how the NHIM persists. Note that this example exhibits a Shil'nikov bifurcation, see Remark 1.15 below.

Example 1.13 (*Perturbation of a compact non-injectively immersed NHIM*) We consider on $\mathbb{R}^3$ the vector field

[18] The order of contact is defined as the degree *up to and including* which the Taylor expansions of the objects agree.

[19] The number of preimages must be countable if M is assumed to be second-countable.

$$\dot{x} = \arctan(x^2) + \varepsilon,$$
$$\dot{y} = y,$$
$$\dot{z} = -z$$

and smoothly modify it outside the cylinder $y^2 + z^2 = 1$ such that it flows in the negative x-direction and connects the basin of repulsion of the origin intersected with $x > 0$ to the basin of attraction intersected with $x < 0$. The perturbation parameter ε is initially set to zero.

Note that the x-axis is a NHIM (the arctangent is there to keep the vector field and tangential growth rate bounded). Due to the modification, the two loops in Fig. 1.8 are also NHIMs of this system, both separately and their union. They start from the origin along the positive x-axis, then diverge from it in opposite directions in the xy-plane; once outside the cylinder $y^2 + z^2 = 1$ they start moving into the negative x direction and finally return to the origin approximately along the xz-plane.

We can parametrize their joint image with an injection ι_1 mapping $M = \{0, 1\} \times S^1$ separately onto the two loops, but we can also parametrize with ι_2 that maps $M = S^1$ onto the full figure eight image. If we perturb to $\varepsilon > 0$, then ι_1 will result in Fig. 1.9 where the two loops are separated, while ι_2 will result in Fig. 1.10 which has one loop, but the middle of the figure eight does not intersect anymore. Figure 1.11 shows how the two orbits from the separate loops closely pass the x–axis along hyperbolic

Fig. 1.8 A non-injectively immersed manifold with compact image

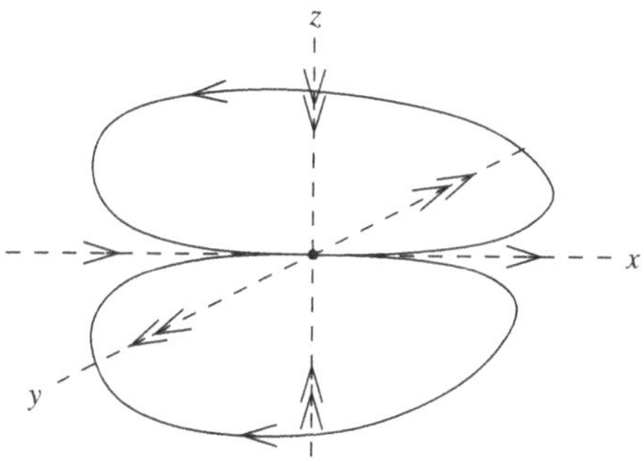

Fig. 1.9 The persistent manifold of ι_1 consisting of two separate loops

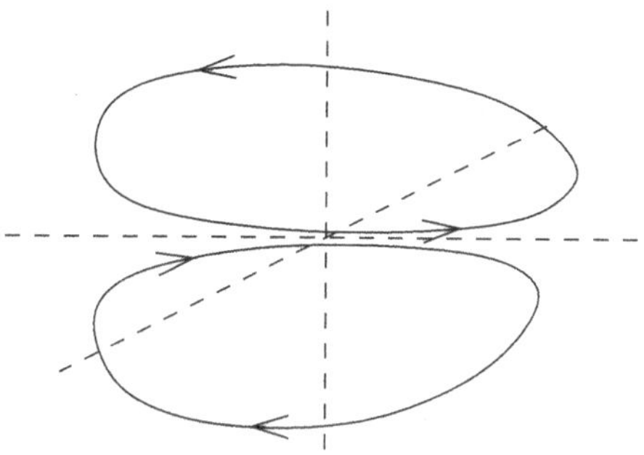

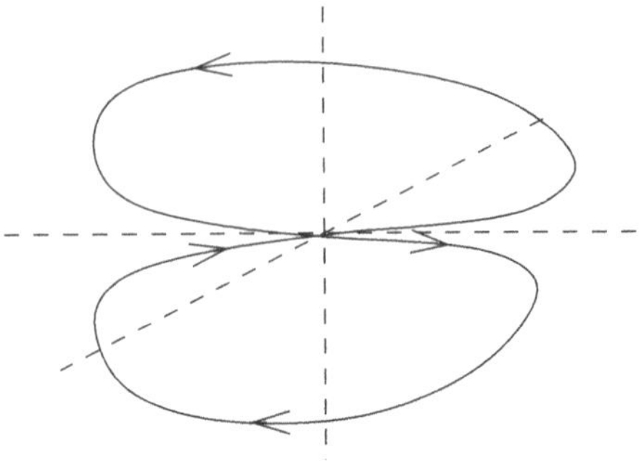

Fig. 1.10 The persistent manifold of ι_2 consisting of one figure eight loop without self-intersection

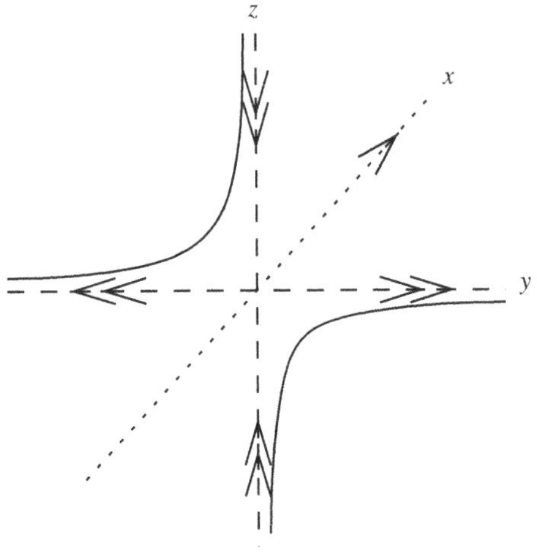

Fig. 1.11 Projection onto the yz-plane showing the orbits of the persistent manifold ι_1 while passing the origin

trajectories. The single orbit of ι_2 follows hyperbolic trajectories through the other two quadrants. ○

Remark 1.14 Note that these different persistent NHIMs do not contradict the uniqueness property of persistence, since the (abstract) manifolds M were different to begin with. Formulated differently, if we consider the universal cover of the tubular neighborhood of $\iota_1(M)$ (deduplicating the origin as image point), then Fig. 1.9 shows the unique invariant manifold that stays in this tubular neighborhood cover. We obtain a different persistent NHIM for any prescribed (possibly infinite) sequence of concatenating the two loops of the original figure eight into an immersion from S^1 (or $\mathbb{R}$ if the sequence is infinite).

Remark 1.15 This example shows a Shil'nikov bifurcation of a saddle-saddle node and provides an alternative proof of the result of Shil'nikov [Šil69, Thm. 3] that for every bi-infinite encoding of the homoclinic loops there exists a unique orbit close to the original homoclinics. We encode the bi-infinite sequence in the immersion ι of a NHIM. For each ι there exists a unique persistent manifold, and these correspond to separate orbits after perturbation. In particular, the periodic orbits correspond to those bi-infinite encodings that are actually periodic and there are countably infinitely many of these. ◇

Finally, we present an example of an injectively immersed (but not embedded) NHIM, see [HPS77, p. 68]. The mapping below is known as Arnold's cat map.

Example 1.16 (*Injectively immersed dense line in the torus*) The matrix

$$A = \begin{pmatrix} 2 & 1 \\ 1 & 1 \end{pmatrix}$$

acting on the two-torus $\mathbb{T}^2$ is an Anosov diffeomorphism. The line through 0 with slope $\frac{1}{2}(1 - \sqrt{5})$ is densely immersed in the torus and it is a NHIM for this discrete system. If we take its suspension, then we have a flow with a NHIM that is densely immersed into the mapping torus $\big([0, 1] \times \mathbb{T}^2\big)/\sim$ with identification $(1, x) \sim (0, A\,x)$. ○

1.6.3 Overflowing Invariant Manifolds

In many applications of normally hyperbolic systems, the manifold M has a boundary ∂M. A typical reason is that the system ceases to be normally hyperbolic across the boundary. This happens, for example, when studying a singularly perturbed, or slow-fast system and in the fast limit there are points on M with zero eigenvalues in the normal direction. At such points, M is not normally hyperbolic anymore, so one must restrict M such that these points are outside of M. Another, somewhat artificial but practical example would be if the invariant manifold is noncompact and one would try to use the classical theorems that are only applicable to compact manifolds by cutting off M to a compact manifold with boundary. One can try to attack this latter case with our more general theory for noncompact manifolds. The additional uniformity assumptions should be checked then.

If M is a manifold with boundary, some persistence results can still be retained. This idea was introduced by Fenichel [Fen72] in studying so-called overflowing invariant manifolds. These are normally hyperbolic manifolds that are invariant under backward time flow, or in other words, only under the forward flow, orbits can leave, i.e. 'overflow' the manifold. The condition of overflowing invariant is slightly stronger: the vector field must strictly point outward at the boundary. This weakened version that the manifold is negatively invariant does come at the additional cost that only stable normal directions are allowed. The time-reversed situation of an inflowing invariant manifold with only unstable normal directions is equivalent. In Sect. 4.3 we discuss how this idea can be incorporated into the Perron method proof.

The attention of the reader is also drawn to the following remark made in [Fen72, p. 214]. If an open submanifold $N \subset M$ is overflowing invariant, and the spectral gap condition is satisfied on N with a higher ratio r_N than on the whole of M, then the persistent manifold $\tilde{N}$ over N retains C^{r_N} smoothness, even if smoothness of $\tilde{M}$ will generally be lower.

1.7 Induced Topology

In this work the topologies for spaces of vector fields, submanifold embeddings, et cetera, are (implicitly) defined by norms and distance functions. The norms we use are uniform C^k-norms for bounded functions, and families with additional exponential growth rates. Let us call the topologies induced by these norms C_b^k-topologies and consider how they compare to two common topologies: the weak and strong Whitney topologies for maps between manifolds, alternatively known as the compact-open and fine topology, see [Hir76].

The weak topology has a subbasis generated by the set of functions g that are close to some function f in C^k-norm on compact subsets in local coordinate charts. This means that for example the function family

$$f_\delta \colon \mathbb{R} \to \mathbb{R} \colon x \mapsto \delta \exp(x^2)$$

converges to zero for $\delta \to 0$ in this topology. On any compact set f_δ will become arbitrarily small when $\delta \to 0$ while it does not converge in uniform norm (nor with additional exponential growth rate). Hence the weak topology is weaker than our induced C_b^k-topologies.

The strong topology has as basis all sets of functions g that are close to some function f on a locally finite cover by compact sets K_i, where g must approximate f in C^k-norm on each K_i in local coordinates up to a given chart-dependent size ε_i. For any function without compact support, a collection $\varepsilon_i > 0$ can be found that converges faster to zero on each larger K_i than the function to zero when $x \to \infty$. Hence the only sequences of functions $\mathbb{R} \to \mathbb{R}$ that converge to the zero function in the strong topology are those with (eventually) compact support. A family f_δ of functions with noncompact support cannot converge to the zero function, as can be seen by using a diagonal argument. The family $f_\delta(x) = \delta \exp(-x^2)$, for example, does not converge to the zero function in the strong topology. Given a locally finite cover of $\mathbb{R}$ by compact sets K_i, we choose $x_i \in K_i$ and corresponding $\varepsilon_i = \exp(-x_i^2)/i$. Then for any given $\delta > 0$, we will have $|f_\delta(x_i)| > \varepsilon_i$ for some large i. On the other hand, this family f_δ obviously converges under the uniform norm with any exponential growth rate. Thus, the strong topology is stronger than our induced C_b^k-topologies, see also the remark in [GG73, p. 43] for noncompact manifolds.

We conclude that the C_b^k-topologies induced by our uniform norms are not equivalent to either the weak or strong Whitney topology, because the weak topology allows arbitrary behavior of functions outside compact sets, while the strong topology completely restricts that behavior. Our norms allow moderate variations at infinity. In general, 'moderate behavior' is not well-defined on a general noncompact manifold, as it depends on the choice of charts. In the setting of bounded geometry, though, the uniform, metric structure makes this behavior unambiguous; we can restrict to normal coordinate charts and consider 'moderate behavior' with respect to these. Note that these topologies are equivalent on compact domains.

1.8 Notation

Here, we will establish some notation and conventions to be used throughout this work. See the index for more specific symbols.

- The letters I and J will denote intervals in $\mathbb{R}$; I will typically represent an interval that is unbounded on one side, while J will be bounded.
- $\varepsilon, \delta > 0$ will denote (small) bounds for continuity-like estimates; $C > 0$ will denote arbitrary bounds. The specific meaning of these symbols will vary depending on context. $\varepsilon_f(\delta)$ will denote a uniform continuity modulus of the function f, that is, $\varepsilon_f : \mathbb{R}_{\geq 0} \to \mathbb{R}_{\geq 0}$ satisfies

$$d(f(x_2), f(x_1)) < \varepsilon_f(d(x_2, x_1)) \quad \text{and} \quad \lim_{\delta \to 0} \varepsilon_f(\delta) = 0. \tag{1.13}$$

  Without subscript f this will denote an arbitrary continuity modulus.

- The D denotes a total derivative, while D_i with index $i \in \mathbb{N}$ denotes a partial derivative with respect to the i-th argument, or, when a subscript symbol is appended, say D_x, then this denotes a partial derivative with respect to the argument commonly referred to by that symbol.
- We use the following symbols to denote classes of function spaces:

C_b	bounded, continuous functions;
$C_{b,u}$	bounded, uniformly continuous functions;
C^k	k times continuously differentiable functions;
$C^{k,\alpha}$	C^k functions with α-Hölder continuous k-th derivative. We will conventionally write $r = k + \alpha \in \mathbb{R}_{\geq 1}$; the Hölder estimates are assumed to be uniform in $C^{k,\alpha}_{b,u}$ spaces.
$\mathcal{L}$	continuous, i.e. bounded, (multi)linear operators;
$\mathfrak{X}$	vector fields;
Γ	sections of a fiber bundle.

Unless otherwise specified, C^k_b and $C^{k,\alpha}_b$ spaces will be endowed with the canonical norms that turn these into Banach spaces, that is,

$$\|f\|_{k,\alpha} = \sum_{0 \leq n \leq k} \sup_x \left\|\mathrm{D}^n f(x)\right\| + \sup_{x_2 \neq x_1} \frac{\left\|\mathrm{D}^k f(x_2) - \mathrm{D}^k f(x_1)\right\|}{d(x_2, x_1)^\alpha}. \tag{1.14}$$

We define the operator norm on a multilinear operator $A \in \mathcal{L}^k(V_1 \times \cdots \times V_k; W)$ as

$$\|A\| = \sup_{\substack{v_i \in V_i \\ \|v_i\| = 1}} \|A(v_1, \ldots, v_k)\|. \tag{1.15}$$

This multilinear operator norm can be extended to sections s of real-valued tensor bundles by taking the operator norm pointwise of $s(x)$ as a multilinear operator into $\mathbb{R}$.

- On a Riemannian manifold, Γ will denote the Christoffel symbols, while Π will be used for parallel transport along a curve given as argument, for example, $\Pi(\gamma|_a^b)$ will denote parallel transport along the curve γ restricted to the interval $[a, b]$. We shall denote induced parallel transport on products of the tangent bundle by $\Pi(\gamma|_a^b)^{\otimes k}$.
- We shall often work with maps that are defined on the tangent space over a point $x \in M$ and denote this dependence on x by a subscript, for example $h_x : \mathrm{T}_x M \to \mathrm{T}_x M$. If we want to refer to the whole family of such maps for all $x \in M$, then we denote this by
$$h_\bullet : \mathrm{T}_\bullet M \to \mathrm{T}_\bullet M,$$
particularly if we want to stress that this family satisfies some properties uniformly in x.
- We use the notation $B(x; \delta)$ not only to indicate open balls of radius δ around a single point x, but also $B(M; \delta)$ to indicate a (tubular) neighborhood of some set or submanifold M, that is,
$$B(M; \delta) = \{x | d(M, x) < \delta\}.$$

The following definition of a scale of Banach spaces (cf. [VG87]) is fundamental to the rest of this work.

Definition 1.17 *Let X be a normed linear space and $\mathcal{F} = C(I; X)$ the space of continuous functions from an interval $I \subset \mathbb{R}$ to X. We define a family of exponential growth norms with parameter $\rho \in \mathbb{R}$ by*

$$\|f\|_\rho = \sup_{t \in I} \|f(t)\| \, e^{-\rho t} \quad \textit{for} \quad f \in \mathcal{F}. \tag{1.16}$$

We define $B^\rho(I; X)$ to be the normed space consisting of all functions $f \in \mathcal{F}$ with $\|f\|_\rho < \infty$. If X is a Banach space, then $B^\rho(I; X)$ is a Banach space as well.

Remark 1.18 When the interval I is bounded from below, then the embedding $B^{\rho_1}(I; X) \hookrightarrow B^{\rho_2}(I; X)$ is continuous for $\rho_1 \le \rho_2$. The time reversed version when I is bounded above and $\rho_2 \le \rho_1$ holds, will frequently recur throughout this work. See also Remark B.4 and the note on integrals of exponentials (1.18) below. In Chap. 3 we shall use $I = \mathbb{R}_{\le 0}$ and negative rates ρ, while in the appendices B and C we use (the somewhat more natural) $I = \mathbb{R}_{\ge 0}$; though ρ's can take both signs there. ◇

The definition of an exponential growth norm can be generalized to curves mapping into a metric space. Let (X, d) be a metric space, then analogously to (1.17), we define a family of exponential growth distance functions on $\mathcal{F}$ by

$$d_\rho(f_1, f_2) = \sup_{t \in I} d\big(f_1(t), f_2(t)\big)\, e^{-\rho t}. \tag{1.17}$$

Note that this distance function might be infinite for some $x_1, x_2 \in \mathcal{F}$.

We will be working with exponential growth estimates of the form $C\, e^{\rho t}$ throughout this paper. The pair of numbers $C > 0$, $\rho \in \mathbb{R}$ that determine such a growth estimate will be referred to as *exponential growth numbers*, and ρ as an *exponential growth rate*.

We will frequently encounter integrals over a time interval, where the integrand obeys an exponential estimate. As long as the interval $[a, b]$ is bounded in the direction of exponential growth and $\rho \neq 0$, these can be estimated as

$$\int_a^b e^{\rho t}\, \mathrm{d}t \leq \frac{1}{|\rho|} \exp\Big(\sup_{t \in [a,b]} \rho\, t \Big). \tag{1.18}$$

We also state here some basic facts about uniformly Hölder continuous functions.

Lemma 1.19 (Product rule for Hölder continuity) *Let $f, g \in C^\alpha_{b,u}$ be defined on spaces such that the product $f \cdot g$ is well-defined. Then also $f \cdot g \in C^\alpha_{b,u}$.*

Proof Let $\|f\|_0, \|g\|_0 \leq M$ and let $C_{f,\alpha}, C_{g,\alpha}$ be the respective Hölder coefficients of f, g. Then we have for all $x_1 \neq x_2$

$$\begin{aligned}\|f(x_2)\, g(x_2) - f(x_1)\, g(x_1)\| &\leq \|f(x_2)\|\, \|g(x_2) - g(x_1)\| + \|f(x_2) - f(x_2)\|\, \|g(x_2)\| \\ &\leq M\, (C_{f,\alpha} + C_{g,\alpha})\, \|x - y\|^\alpha,\end{aligned}$$

which exhibits the Hölder coefficient $M\,(C_{f,\alpha} + C_{g,\alpha})$ for the product, and $f \cdot g$ is clearly bounded by M^2. □

Lemma 1.20 *Let $f \in C^\alpha_{b,u}$. Then it also holds that $f \in C^\beta_{b,u}$ for any $0 < \beta < \alpha$.*

Proof Let M be the bound on f, and C_α its α-Hölder coefficient. For $\|x_2 - x_1\| \leq 1$ the estimate for β follows automatically from that of α. For $\|x_2 - x_1\| > 1$ we use boundedness to obtain

$$\|f(x_2) - f(x_1)\| \leq 2\, M \leq 2\, M\, \|x_2 - x_1\|^\beta.$$

Hence, $C_\beta = \max(C_\alpha, 2\, M)$ suffices as β-Hölder coefficient. □

Typographical Conventions

As usual we close proofs with the symbol □, while we shall use ◊ and ○ to denote the end of (a series of) remarks or examples, respectively.

Chapter 2
Manifolds of Bounded Geometry

For noncompact normally hyperbolic systems, uniformity assumptions that were implicit in the compact case must be made explicit. Not only assumptions on the vector field, but on the underlying space as well. For this we need the concept of bounded geometry; Sect. 3.3 contains a discussion and examples for why we require this concept.

The class of manifolds of bounded geometry allows us to uniformly apply constructions that are well-known for compact manifolds. We single out the atlas of normal coordinate charts and derive from the very definition of bounded geometry that all constructions and estimates are uniform over all such charts. For completeness, we present here all results that we need later on. Some of these results are already present in the literature: the construction of a uniformly locally finite cover and a subordinate C^k uniformly bounded partition of unity, and bounded coordinate transformations can be found in [Shu92; Sch01], for example, while [Roe88] includes the result on finite coloring of the connectedness graph of a uniformly locally finite cover and the construction of a trivial bundle embedding in Proposition 7.5 with a sketch of the proof. I have not been able to find in the literature the results about the existence of a uniform tubular neighborhood and the approximation of a submanifold by a smoothed manifold. Submanifolds are allowed to be non-injectively immersed.

This chapter is organized as follows. First, the material is presented that is already required for the global coordinate setting of Theorem 3.2. These include the basic definitions of bounded geometry, related results on bounded coordinate transition maps, uniform covers and partitions of unity, and an explicit relation between holonomy and curvature. Then we continue to work towards the final goal of this chapter: to reduce a noncompact normally hyperbolic system from a setting in general manifolds to a trivial bundle setting, in order to generalize the persistence theorem to the former setting. To this end, we need some more technical results: a uniform tubular neighborhood, smooth approximation of a submanifold, and embedding into a trivial bundle.

This chapter relies heavily on some more advanced concepts from differential and specifically Riemannian geometry. On the other hand, the results are used as

J. Eldering, *Normally Hyperbolic Invariant Manifolds*, Atlantis Series in Dynamical Systems 2, DOI: 10.2991/978-94-6239-003-4_2,

tools in solving a dynamical systems problem. Appendix F provides a quick review for non-experts of the most relevant geometric concepts used here. It also provides further references to the literature. We shall assume the contents of this appendix known from here on.

I suggest the reader to at least take a glance at the first two sections of this chapter to familiarize himself with the basic definitions and results of bounded geometry, without the need to go through the details of the proofs. Then, depending on his interest, he can choose to delve into the more technical geometric details or skip to Chap. 3 for the more analytical side of the proof of Theorem 3.2, and possibly return later to read how Theorem 3.1 is reduced to the former.

2.1 Bounded Geometry

We follow the definition in [Eic91] to introduce bounded geometry. Recall that the injectivity radius $r_{\mathrm{inj}}(x)$ at a point $x \in M$ is the maximum radius for which the exponential map at x is a diffeomorphism, see also Appendix F.

Definition 2.1 (Bounded geometry) *We say that a complete, finite-dimensional Riemannian manifold (M, g) has k-th order bounded geometry when the following conditions are satisfied:*

(I) *the global injectivity radius $r_{inj}(M) = \inf_{x \in M} r_{inj}(x)$ is positive, $r_{inj}(M) > 0$;*

($\mathbf{B_k}$) *the Riemannian curvature R and its covariant derivatives up to k-th order are uniformly bounded,*

$$\forall\, 0 \le i \le k : \sup_{x \in M} \|\nabla^i R(x)\| < \infty,$$

with operator norm of $\nabla^i R(x)$ as an element of the tensor bundle over $x \in M$.

Remark 2.2 The conditions (I) and (B_k) are independent. We present a simple example which exhibits zero infimum for the injectivity radius while all derivatives of the curvature are globally bounded. Indeed, let $M = \mathbb{R} \times S^1$ be a cylinder with metric $g = \mathrm{d}x^2 + e^{-2x}\,\mathrm{d}\theta^2$ in coordinates (x, θ), see also Fig. 3.3 on page 88.[1] The injectivity radius $r_{\mathrm{inj}}(M)$ is zero since the cylinder circumference shrinks to zero with $x \to \infty$. Global boundedness of the curvature and all of its derivatives follows from a symmetry argument. The family

$$\varphi_{\xi,\alpha} : (x, \theta) \mapsto (x + \xi, e^{\xi}\,\theta + \alpha) \qquad \text{with} \quad \xi \in \mathbb{R},\ \alpha \in [0, 2\pi)$$

[1] This is a noncompact surface with constant negative curvature, hence it cannot be isometrically embedded into $\mathbb{R}^3$, see [Hil01]. The embedding is nearly isometric for $x \gg 0$ though, so the figure is still a good representation there.

is a set of local isomorphisms that acts transitively on M. That is, for any two points $(x_1, \theta_1), (x_2, \theta_2) \in M$ there exist ξ, α and a neighborhood $U \ni (x_1, \theta_1)$ such that $\varphi_{\xi,\alpha} : U \to \varphi_{\xi,\alpha}(U)$ is an isomorphism and $\varphi_{\xi,\alpha}(x_1, \theta_1) = (x_2, \theta_2)$. For any $(x, \theta) \in U$ and $v, w \in \mathrm{T}_{(x,\theta)}M$ we have

$$\begin{aligned}(\varphi_{\xi,\alpha}^* g)_{(x,\theta)}(v, w) &= g_{(x+\xi, e^\xi \theta+\alpha)}\big(\mathrm{D}\varphi_{\xi,\alpha}(x,\theta)\, v, \mathrm{D}\varphi_{\xi,\alpha}(x,\theta)\, w\big)\\ &= \mathrm{d}x(v)\mathrm{d}x(w) + e^{-2(x+\xi)}\, e^\xi \mathrm{d}\theta(v)\, e^\xi \mathrm{d}\theta(w)\\ &= g_{(x,\theta)}(v, w)\end{aligned}$$

so $\varphi_{\xi,\alpha}^* g = g$ on U. Since the curvature and its derivatives are locally determined, this implies that these are constant across M, hence uniformly bounded (actually all derivatives of R vanish). Note that these local isometries do not imply a finite global injectivity radius since the size of the neighborhood U does depend on the points $(x_1, \theta_1), (x_2, \theta_2) \in M$. ◇

Example 2.3 (*Manifolds of bounded geometry*) The following are examples of manifolds with bounded geometry of any (i.e. infinite) order.

- Euclidean space with the standard metric trivially has bounded geometry.
- A smooth, compact Riemannian manifold M has bounded geometry as well; both the injectivity radius and the curvature including derivatives are continuous functions, so these attain their finite minimum and maxima, respectively, on M. If $M \in C^{k+2}$, then it has bounded geometry of order k.
- Noncompact, smooth Riemannian manifolds that possess a transitive group of isomorphisms (such as hyperbolic space) have bounded geometry since the finite injectivity radius and curvature estimates at any single point translate to a uniform estimate for all points under isomorphisms. Note that the example in Remark 2.2 above shows that it is not sufficient to have local isometries.

More manifolds of bounded geometry can be constructed with these basic building blocks in the following ways.

- The product of a finite number of manifolds of bounded geometry again has bounded geometry, since the direct sum structure of the metric is inherited by the exponential map and curvature. We give an outline of the proof. In a product coordinate chart
$$(\varphi_1, \varphi_2) : U_1 \times U_2 \to \mathbb{R}^{n_1} \times \mathbb{R}^{n_2}$$
with coordinates (x_1, x_2), the metric has diagonal form
$$g(x_1, x_2) = (g_1 \oplus g_2)(x_1, x_2) = \begin{pmatrix} g_1(x_1) & 0 \\ 0 & g_2(x_2) \end{pmatrix}.$$
The coordinate dependence on x_1, x_2 is non-mixed and this is preserved under taking derivatives and index contractions, so R will split into a direct sum of R_1 and R_2 again. This can be extended to derivatives of R.

A geodesic in $M_1 \times M_2$ is precisely given by $\gamma = (\gamma_1, \gamma_2)$ where γ_1, γ_2 are geodesics parametrized with constant speed in M_1, M_2, respectively. This follows easily since minimization of length is equivalent to minimization of the energy functional

$$2\,E(\gamma) = \int_a^b g(\dot\gamma, \dot\gamma)\mathrm{d}t = \int_a^b g_1(\dot\gamma_1, \dot\gamma_1) + g_2(\dot\gamma_2, \dot\gamma_2)\mathrm{d}t$$

and this splits nicely into independent minimization problems for γ_1 and γ_2. With a little effort one sees that $r_{\mathrm{inj}}(M) \geq \min\big(r_{\mathrm{inj}}(M_1), r_{\mathrm{inj}}(M_2)\big)$.

- If we take a finite connected sum of manifolds with bounded geometry such that the gluing modifications are smooth and contained in a compact set, then the resulting manifold has bounded geometry again.
- We can endow the tangent bundle $\mathrm{T}M$ of a Riemannian manifold (M, g) with the natural Sasaki metric [Sas58]. Let x^i denote coordinates on an open neighborhood $U \subset M$. These coordinate functions can be pulled back to $\mathrm{T}U$ and the one-forms $\mathrm{d}x^i$ can be viewed as additional coordinates v^i such that the x^i, v^j together form a complete set of induced coordinates on $\mathrm{T}U$. With respect to these coordinates the Sasaki metric is given by

$$\hat g(x, v) = g_{ij}(x)\big(\mathrm{d}x^i\,\mathrm{d}x^j + \mathrm{D}v^i\,\mathrm{D}v^j\big) \qquad \text{where} \quad \mathrm{D}v^i = \mathrm{d}v^i + \Gamma^i_{jk} v^j\,\mathrm{d}x^k \tag{2.1}$$

and Γ^i_{jk} denote the Christoffel symbols on M, while the $\mathrm{d}v^i$ are one-forms on the manifold $\mathrm{T}U$.
Bounded geometry of (M, g) is not inherited by $(\mathrm{T}M, \hat g)$ since the extended Riemannian curvature $\hat R$ contains unbounded terms v when expressed in terms of R, see [GK02, Prop. 7.5]. These expressions do readily show that the restriction $\mathrm{T}^r M = \big\{(x, v) \in \mathrm{T}M \mid g_x(v, v) \leq r^2\big\}$ satisfies curvature bounds of order $k - 1$ if (M, g) has k-bounded geometry. The geodesic flow equation is given in induced coordinates by [Sas58, Eq. (7.7)]. By application of Theorem A.6, one can then show that the injectivity radius is bounded.
Note that $\mathrm{T}^r M$ is a manifold with boundary, but this is not problematic in our setting as long as the invariant submanifold stays away from the boundary. Alternatively one could try to use results from [Sch01]. ○

When we say that a manifold has bounded geometry without specifying the order k, then it is assumed that the order is infinite, $k = \infty$, or sufficiently large. When $k \geq 1$ we have the following result, see [Eic91, Thm. 2.4 and Cor. 2.5]. In case $k = \infty$ the converse also holds [Roe88, Prop. 2.4].

Theorem 2.4 (Boundedness of the metric) *Let (M, g) be a Riemannian manifold of k-bounded geometry. Then there exists a $\delta > 0$ such that the metric up to its k-th order derivatives and the Christoffel symbols up to its $(k-1)$-th order derivatives are bounded in normal coordinates of radius δ around each $x in M$, and the bounds are uniform in x.*

This basic fact can be used to make the properties of all kinds of constructions uniform over a noncompact manifold. Note that here and in the following, all uniformity estimates are assumed globally valid, that is, independent of the point $x \in M$. To stress this, we shall use notation $f_\bullet$, for example as in Definition 2.9, to indicate that the family of maps $\{f_x\}_{x\in M}$ satisfies continuity estimates independent of x

With Theorem 2.4 at hand, we shall exclusively use normal coordinates for local coordinate calculations. To establish notation, we say that

$$\varphi = \exp_x{}^{-1} : B(x;\delta) \subset M \to B(0;\delta) \subset \mathrm{T}_x M \tag{2.2}$$

is a normal coordinate chart at $x \in M$. The radius δ will always be chosen smaller than the injectivity radius $r_{\mathrm{inj}}(M)$, so φ is a diffeomorphism. Each tangent space $\mathrm{T}_x M$ carries the inner product g_x, hence is isometric to Euclidean space $\mathbb{R}^n$ (but identification requires a choice of basis).

Proposition 2.5 *Let (M, g) be a Riemannian manifold of $k \geq 1$ bounded geometry. For every $C > 1$ there exists a $\delta > 0$ such that the normal coordinate charts φ_x in (2.2) are defined on $B(x;\delta)$ for each $x \in M$ and the Euclidean distance d_E on the normal coordinates is uniformly C-equivalent to the metric distance d induced by M, that is,*

$$\forall x_1, x_2 \in B(x;\delta): \quad C^{-1}\, d(x_1, x_2) \leq d_E(\varphi_x(x_1), \varphi_x(x_2)) \leq C\, d(x_1, x_2).$$

Proof Let $\delta < \frac{1}{2} r_{\mathrm{inj}}(M)$ and $x \in M$. We consider a normal coordinate chart φ_x on $B(x; 2\delta)$. According to Theorem 2.4, the metric g and its derivatives are bounded in normal coordinates. We have $(\exp_x^* g)(0) = g_x$, the Euclidean inner product on $\mathrm{T}_x M$, while the total derivative $\mathrm{D}(\exp_x^* g)(\xi)$ is bounded on $B(0; 2\delta) \ni \xi$, say by $\|\mathrm{D}(\exp_x^* g)(\xi)\| \leq C_1$, independent of $x \in M$. By the mean value theorem this induces the uniform bounds

$$1 - 2\delta\, C_1 \leq \|(\exp_x^* g)(\xi)\| \leq 1 + 2\delta\, C_1.$$

Let $x_1, x_2 \in B(x;\delta)$ and let γ_E be the straight curve between $\varphi_x(x_1)$ and $\varphi_x(x_2)$ in $\mathrm{T}_x M$ parametrized by arc length. This curve γ_E attains the Euclidean distance $l_E(\gamma_E) = d_E(\varphi_x(x_1), \varphi_x(x_2))$. On the other hand, it gives an upper bound on the metric distance

$$d(x_1, x_2) = \inf_{\gamma} l(\gamma) \le \int_0^{l_E(\gamma_E)} \sqrt{(\exp_x^* g)_{\gamma_E(t)}\big(\gamma'_E(t), \gamma'_E(t)\big)}\mathrm{d}t$$
$$\le \sqrt{1 + 2\delta\, C_1}\; d_E\big(\varphi_x(x_1), \varphi_x(x_2)\big).$$

Let γ be a geodesic minimizing the distance $d(x_1, x_2)$. Then γ is contained in $B(x; 2\delta)$: the distance from each x_i to the boundary of $B(x; 2\delta)$ is at least δ, so if γ would leave and re-enter $B(x; 2\delta)$ then its length would be at least 2δ. On the other hand, x_1 and x_2 can be connected via x with a curve of length less than 2δ. Let us write $\eta = \varphi_x \circ \gamma$ and assume that η is parametrized by arc length with respect to the Euclidean metric g_x. Then we obtain an inverse estimate to the one above:

$$d_E(\varphi_x(x_1), \varphi_x(x_2)) \le \int_0^{l_E(\eta)} 1\mathrm{d}t \le \int_0^{l_E(\eta)} (1 - 2\delta\, C_1)^{-\frac{1}{2}} \sqrt{(\exp_x^* g)_{\eta(t)}\big(\eta'(t), \eta'(t)\big)}\mathrm{d}t$$
$$= (1 - 2\delta\, C_1)^{-\frac{1}{2}} \int_0^{l_E(\eta)} \sqrt{g_{\gamma(t)}\big(\gamma'(t), \gamma'(t)\big)}\mathrm{d}t$$
$$\le (1 - 2\delta\, C_1)^{-\frac{1}{2}}\, d(x_1, x_2).$$

Finally, we complete the proof by choosing $\delta > 0$ small enough that

$$\max\big((1 + 2\delta\, C_1)^{\frac{1}{2}}, (1 - 2\delta\, C_1)^{-\frac{1}{2}}\big) \le C. \qquad \square$$

From here on, we shall frequently represent objects living in $B(x; \delta) \subset M$ on normal coordinate neighborhoods $B(0; \delta) \subset \mathrm{T}_x M$ via the normal coordinate chart φ_x. We will mostly use $B(x; \delta)$ to clearly indicate the base point, or $B(0_x; \delta) \subset \mathrm{T}_x M$ to stress the tangent space domain of the coordinates as well. In spaces of bounded geometry, normal coordinate charts are the natural charts to works in and coordinate transition maps are not just smooth, but uniformly bounded, as stated in the following lemma.

Lemma 2.6 (Boundedness of transition maps) *Let (M, g) be a Riemannian manifold of k-bounded geometry with $k \ge 2$. There exists a δ with $0 < \delta < r_{inj}(M)$ and constants $C,\ L > 0$ such that for all $x_1, x_2 \in M$ with $d(x_1, x_2) < \delta$ the following holds.*

1. *The coordinate transition map*

$$\varphi_{2,1} = \varphi_2 \circ \varphi_1^{-1} : U \to \mathrm{T}_{x_2} M \qquad \textit{with} \quad U = \varphi_1(B(x_1; \delta) \cap B(x_2; \delta)) \subset \mathrm{T}_{x_1} M \tag{2.3}$$

   is C^{k-1} bounded with $\|\varphi_{2,1}\|_{k-1} \le C$.
2. *Let $\gamma_{2,1} : [0, 1] \to B(x_1; \delta)$ be the unique shortest geodesic connecting x_1 and x_2 and let $\Pi(\gamma_{2,1})$ be the associated parallel transport. Then the map*

$$\varphi_{2,1} - \Pi(\gamma_{2,1}) : U \to \mathrm{T}_{x_2} M$$

   has C^{k-2}-norm bounded by the Lipschitz estimate

$$\|\varphi_{2,1} - \Pi(\gamma_{2,1})\|_{k-2} \leq L\, d(x_1, x_2). \tag{2.4}$$

Remark 2.7 One degree of smoothness is lost because the exponential map is defined in terms of the geodesic flow. This flow in turn is defined in terms of the Christoffel symbols, which depend on derivatives of the metric, so these are only C^{k-1} bounded. We lose another degree of smoothness in estimating $\varphi_{2,1} - \Pi(\gamma_{2,1})$ since the Lipschitz estimate follows from a uniform bound on one higher derivative of these. ◇

We shall first compare both $\varphi_{2,1}$ and $\Pi(\gamma_{2,1})$ to the identity in normal coordinates and finally conclude with the triangle inequality that their difference must be small. We compare $\varphi_{2,1}$ to the parallel transport $\Pi(\gamma_{2,1})$ since this is the most natural way to identify the tangent spaces $\mathrm{T}_{x_1}M$ and $\mathrm{T}_{x_2}M$.

Proof Let $B(x_1; \delta)$, $B(x_2; \delta)$ be two normal coordinate neighborhoods with non-empty intersection. The coordinate transition map $\varphi_{2,1} = \varphi_2 \circ \varphi_1^{-1} = \exp_{x_2}^{-1} \circ \exp_{x_1}$ can be studied as the exponential map $\exp_{x_1} : \mathrm{T}_{x_1}M \to M$ in normal coordinates on $B(x_2; \delta)$, since $\varphi_2 = \exp_{x_2}^{-1}$. From here on, we will implicitly be working in normal coordinates around x_2, using some choice of basis to isometrically identify $\mathrm{T}_{x_2}M \cong \mathbb{R}^n$.

Let $x \in B(x_1; \delta) \cap B(x_2; \delta)$, hence $x_1 \in B(x_2; 2\,\delta)$. We choose $\delta \leq 1$, and small enough so that the results of Theorem 2.4 and Proposition 2.5 (with $C = 2$) hold for $2\,\delta$. The exponential map is given by the time-one geodesic flow projected on the base manifold. For the base point x_2, this is the identity map, while for the base point x_1 we will show that it is a small perturbation thereof. The geodesic flow on $\mathrm{T}M$ is given in local coordinates by

$$\begin{aligned} \dot{x}^i &= v^i, \\ \dot{v}^i &= -\Gamma^i{}_{jk}(x)\, v^j\, v^k, \end{aligned} \tag{2.5}$$

where Γ^i_{jk} denote the Christoffel symbols with respect to the coordinates x^i on M and the v^j are induced additional coordinates on $\mathrm{T}M$, see the explanation above (2.1). The Christoffel symbols are C^{k-1} bounded due to Theorem 2.4. Let Υ^t denote the geodesic flow of (2.5) on $\mathrm{T}M$ restricted to $B(x_2; 2\,\delta)$. We denote by $(x(t), v(t))$ a solution curve of Υ^t. The geodesic flow preserves the length of tangent vectors with respect to the metric g, so we have $\|v(t)\| \leq 2\,\|v(0)\| \leq 2\,\delta$ with respect to the Euclidean distance in the normal coordinates. This implies that the vector field (2.5) is bounded in these induced coordinates. Hence, by Theorem A.6, $\Upsilon^t \in C_b^{k-1}$ is bounded as well on the interval $[0, 1]$. Moreover, $\mathrm{D}\Upsilon^t \in C_b^{k-2}$ exhibits a Lipschitz estimate for the base point dependence $\|\varphi_2(x_1)\|_E$. By Proposition 2.5 the local Euclidean distance is equivalent to the distance on M, so $\|\varphi_2(x_1)\|_E \leq 2\,d(x_1, x_2)$. These conclusions directly translate to $\exp_x(\,\cdot\,) = \pi \circ \Upsilon^1(x, \,\cdot\,)$ and we conclude that $\varphi_{2,1} = \exp_{x_2}^{-1} \circ \exp_{x_1} \in C_b^{k-1}$ with bound $C > 0$ uniform in $x_1, x_2 \in M$ and $\|\varphi_{2,1} - \mathbb{1}\|_{k-2} \leq L'\,d(x_1, x_2)$ for some $L' > 0$.

The parallel transport $\Pi(\gamma_{2,1})$ is given by integrating the pullback of the connection along $\gamma_{2,1}$. This yields a differential equation similar to (2.5) and similarly leads to C^{k-1} boundedness estimates in normal coordinates and Lipschitz estimates for the C^{k-2}-norm. Thus, the difference $\varphi_{2,1} - \Pi(\gamma_{2,1})$ is C^{k-1} bounded, and has C^{k-2}-norm that satisfies the Lipschitz estimate (2.4) for some $L > 0$. □

Definition 2.8 **(M-small coordinate radius)** *Let (M, g) be a Riemannian manifold of bounded geometry. We define $\delta > 0$ to be M-*small *if Theorem 2.4 and Lemma 2.6 hold on all normal coordinate charts of radius δ.*

Note that such a $\delta > 0$ always exists. From now on, we shall always assume to have selected such a δ for any given manifold of bounded geometry and restrict its atlas to include these normal coordinate charts only.

Lemma 2.6 shows that normal coordinate transformations respect C^k boundedness of functions in coordinate representations. Thus, it is natural to consider manifolds of bounded geometry as the class of C^k bounded manifolds with respect to this restricted atlas. This also makes the following definition natural.

Definition 2.9 **($\mathbf{C^k}$ bounded maps)** *Let X, Y be Riemannian manifolds of $k+1$-bounded geometry and $f \in C^k(X; Y)$. We say that f is of class C^k_b when there exist X, Y-small $\delta_X, \delta_Y > 0$ such that for each $x \in X$ we have $f(B(x; \delta_X)) \subset B(f(x); \delta_Y)$ and the representation*

$$\tilde{f}_x = \exp^{-1}_{f(x)} \circ f \circ \exp_x : B(0; \delta_X) \subset \mathrm{T}_x X \to \mathrm{T}_y Y \tag{2.6}$$

in normal coordinates is of class C^k_b and the associated C^k-norms of $\tilde{f}_\bullet$ are bounded uniformly in $x \in X$. We define the classes of $C^k_{b,u}(X; Y)$ and $C^{k,\alpha}_{b,u}(X; Y)$ functions analogously when X, Y are of $k+2$-bounded geometry.

Remark 2.10 We shall say that a vector field $v \in \mathfrak{X}(X)$ is of class C^k_b, also denoted by $v \in \mathfrak{X}^k_b(X)$, when $v \in C^k_b$ with respect to coordinates on $\mathrm{T}M$ induced by normal coordinates on M. This is slightly different from normal coordinates on $\mathrm{T}M$ induced by the metric (2.1). Note that since $\|v\| \le r$ is assumed bounded, we could restrict to the submanifold $\mathrm{T}^r M$ of bounded geometry and consider $v \in C^k_b(M; \mathrm{T}^r M)$, but this is less practical.

Remark 2.11 The manifolds X, Y need to have bounded geometry of one or two degrees higher than the smoothness of the maps to preserve boundedness and uniform continuity estimates under normal coordinate transformations. This shall from now on always be an implicit assumption.

Remark 2.12 *(Locally/globally defined continuity modulus)* The continuity modulus ε_f of a function $f \in C^k_{b,u}(X; Y)$ is only defined on the interval $[0, \delta_X) \subset \mathbb{R}$. On the other hand, $\|\mathrm{D}^k f(x)\|$ is globally well-defined in terms local charts and assumed to be bounded. We shall want to compare $\mathrm{D}^k f$ at points x_1, x_2 far apart. If we have isometric isomorphisms

$$\varphi : \mathrm{T}_{x_1} X \xrightarrow{\sim} \mathrm{T}_{x_2} X \quad \text{and} \quad \psi : \mathrm{T}_{f(x_1)} Y \xrightarrow{\sim} \mathrm{T}_{f(x_2)} Y,$$

then this allows us to compare

$$\|\mathrm{D}^k f(x_2) \circ \varphi^{\otimes k} - \psi \circ \mathrm{D}^k f(x_1)\| \leq \|\mathrm{D}^k f(x_2)\| + \|\mathrm{D}^k f(x_1)\|. \tag{2.7}$$

Note that the right-hand expression does not depend on the choice[2] of isomorphisms.

Thus, with such isomorphisms at hand, we can use (2.7) to *heuristically* extend the local to a global continuity modulus. That is, for nearby points x_1, x_2 we use an estimate in terms of local charts; if this is not possible, then the points must be separated by a distance larger than a δ as in Definition 2.8. Since the functions we consider are globally bounded, we then use some (non-canonical) choice to identify the vector bundle fibers over x_1, x_2 that the function lives in and estimate by the right-hand side of (2.7). This estimate is crude but independent of the choice of identification and will always satisfy our needs. For example, if $f \in C^{k,\alpha}_{b,u}(X; Y)$, with Hölder coefficient C_α locally for $d(x_1, x_2) \leq \delta$ then we have

$$\|\mathrm{D}^k f(x_2) - \mathrm{D}^k f(x_1)\| \leq \begin{cases} C_\alpha\, d(x_1, x_2)^\alpha & \text{if } d(x_1, x_2) < \delta, \\ \frac{2\,\|f\|_k}{\delta^\alpha}\, d(x_1, x_2)^\alpha & \text{else.} \end{cases}$$

This shows that we can heuristically consider $\max\big(C_\alpha, \frac{2\,\|f\|_k}{\delta^\alpha}\big)$ as a global Hölder coefficient. $\diamond$

The following proposition shows that we may measure continuity of the derivatives of a function f using local parallel transport. With the remark above we see how it can be extended to a global continuity modulus if a (non-unique) choice is made for how to connect non-close points x_1, x_2 by a path; this idea will be developed in Sect. 3.7.4.

Proposition 2.13 (Equivalence of continuity moduli) *Let X, Y be Riemannian manifolds of bounded geometry and $f \in C^k_b(X; Y)$. Then the following statements are equivalent:*

1. *$f \in C^{k,\alpha}_{b,u}(X; Y)$ according to Definition 2.9;*
2. *we have the continuity estimate*

$$\begin{aligned} &\exists\, \varepsilon_{f,\Pi} \in C^\alpha(\mathbb{R}_+; \mathbb{R}_+),\ \delta_0 > 0 : \forall\, x_1, x_2 \in X,\ d(x_1, x_2) \leq \delta_0 : \\ &\qquad \|\mathrm{D}^k \tilde{f}_{x_2}(0) \cdot \Pi(\gamma_{2,1})^{\otimes k} - \Pi(\eta_{2,1}) \cdot \mathrm{D}^k \tilde{f}_{x_1}(0)\| \leq \varepsilon_{f,\Pi}(d(x_1, x_2)), \end{aligned} \tag{2.8}$$

[2] In practice, we shall use isomorphisms defined by parallel transport on $X = Y$, cf. Proposition 2.13. This is a non-canonical choice, since it depends on the path connecting x_1, x_2. A canonical choice that depends continuously on x_1, x_2 cannot be made in general, since it would imply that the tangent bundle is trivializable.

where $\Pi(\eta_{2,1})$ and $\Pi(\gamma_{2,1})^{\otimes k}$ denote parallel transport along the unique shortest geodesic between $f(x_1)$, $f(x_2)$ and x_1, x_2, respectively, and $\varepsilon_{f,\Pi}$ denotes a uniform or α-Hölder continuity modulus.

Proof We first prove the statement in case Y is a normed linear space, hence no parallel transport term $\Pi(\eta_{2,1})$ appears.

Let $\delta_0 \le \delta_X$ as in Definition 2.9 (thus, in particular δ_0 is X-small), and let $d(x_1, x_2) \le \delta_0$. Then we have the Lipschitz estimate $\|\varphi_{2,1} - \Pi(\gamma_{2,1})\| \le L\, d(x_1, x_2)$ while the normal coordinate representations (2.6) of f at x_1, x_2 are related by $\tilde{f}_{x_1} = \tilde{f}_{x_2} \circ \varphi_{2,1}$. This leads to

$$\begin{aligned}
&\|\mathrm{D}^k \tilde{f}_{x_2}(0) \cdot \Pi(\gamma_{2,1})^{\otimes k} - \mathrm{D}^k \tilde{f}_{x_1}(0)\| \\
&\quad = \|\mathrm{D}^k \tilde{f}_{x_2}(0) \cdot \Pi(\gamma_{2,1})^{\otimes k} - \mathrm{D}^k[\tilde{f}_{x_2} \circ \varphi_{2,1}](0)\| \\
&\quad \le \|\mathrm{D}^k \tilde{f}_{x_2}(0) \cdot \Pi(\gamma_{2,1})^{\otimes k} - \mathrm{D}^k \tilde{f}_{x_2}(\varphi_{2,1}(0)) \cdot \big(\mathrm{D}\varphi_{2,1}\big)^{\otimes k}\| \\
&\qquad + \sum_{l=1}^{k-1} \|\mathrm{D}^l \tilde{f}_{x_2}(\varphi_{2,1}(0)) \cdot P_{l,k}\big(\mathrm{D}^\bullet \varphi_{2,1}(0)\big)\| \\
&\quad \le \|\mathrm{D}^k \tilde{f}_{x_2}(0) - \mathrm{D}^k \tilde{f}_{x_2}(\varphi_2(x_1))\| + \|\mathrm{D}^k \tilde{f}_{x_2}(\varphi_2(x_1))\|\, \|\Pi(\gamma_{2,1}) - \mathrm{D}\varphi_{2,1}\|^k \\
&\qquad + \sum_{l=1}^{k-1} \|\mathrm{D}^l \tilde{f}_{x_2}(\varphi_2(x_1))\|\, \|P_{l,k}\big(\mathrm{D}^\bullet \varphi_{2,1}(0)\big)\| \\
&\quad \le \varepsilon_f(d(x_2, x_1)) + \|f\|_k \big(L\, d(x_1, x_2)\big)^k + \sum_{l=1}^{k-1} \|f\|_l\, \|P_{l,k}\big(\mathrm{D}^\bullet \varphi_{2,1}(0)\big)\|,
\end{aligned}$$

where ε_f denotes the continuity modulus of f and its derivatives according to Definition 2.9, and the $P_{l,k}$ denote (l, k)-linear maps according to Proposition C.3. We used the fact that both $\Pi(\gamma_{2,1})$ and $\mathrm{D}\varphi_{2,1}(0)$ act on the k-tensor bundle as a k-tuple of copies. By assumption $\|f\|_k$ is bounded, and to estimate the $P_{l,k}$ terms, we note that $l < k$, so each of the $P_{l,k}$ contains at least a factor $\mathrm{D}^i \varphi_{2,1}(0)$ with $i \ge 2$. Since $\varphi_{2,1}$ is close to $\Pi(\gamma_{2,1})$ and $\mathrm{D}^i \Pi(\gamma_{2,1}) = 0$ for $i \ge 2$, it follows that

$$\|P_{l,k}\big(\mathrm{D}^\bullet \varphi_{2,1}(0)\big)\| \le C\, L\, d(x_1, x_2)$$

for some constant C independent of x_1, x_2. This shows that the continuity modulus $\varepsilon_{f,\Pi}$ of (2.8) can be estimated by the continuity modulus ε_f plus additional Lipschitz terms. We can reverse the estimates above to arrive at the same conclusion when expressing ε_f in terms of $\varepsilon_{f,\Pi}$. Hence, the continuity statements are equivalent for any $\alpha \le 1$.

If Y is a Riemannian manifold of bounded geometry, we just apply the same estimates in the codomain. To this end, we must have $d(f(x_2), f(x_1)) \le \delta_Y$, so we choose δ_0 small enough that

$$d(f(x_2), f(x_1)) \le \|Df\| \, d(x_2, x_1) \le \|f\|_1 \, \delta_0 \le \delta_Y$$

holds with δ_Y as in Definition 2.9. □

The definition of bounded geometry can be extended to vector bundles, see also [Shu92, p 65].

Definition 2.14 **(Vector bundle of bounded geometry)** *Let (M, g) be a manifold of bounded geometry and δ be M-small as in Definition 2.8. We say that a vector bundle $\pi : E \to M$ with fiber F has k-th order bounded geometry when there exist preferred trivializations*

$$\tau : \pi^{-1}\big(B(m; \delta)\big) \to B(m; \delta) \times F \quad \textit{for each } m \in M \tag{2.9}$$

such that if we have a transition function $\varphi_{2,1} = \tau_2 \circ \tau_1{}^{-1}$ between two trivializations on $B(m_1; \delta)$ and $B(m_2; \delta)$, then the function $g : B(m_1; \delta) \cap B(m_2; \delta) \to \mathcal{L}(F)$ defined by $\varphi_{2,1}(m, f) = g(m) f$ satisfies $g \in C^k_b$ independent of the points $m_1, m_2 \in M$.

Remark 2.15 Note that we could have replaced $B(m; \delta)$ by arbitrary (preferred) coordinate charts. The relevant property is that we express uniformity of the transition functions in terms of uniformity of the function g with respect to the underlying coordinate charts of M, which are normal coordinates in our case. ◇

It follows from Lemma 2.6 that the tangent bundle TM has bounded geometry of order $k-2$ if (M, g) has bounded geometry of order $k \ge 2$. One order of smoothness is lost (beyond the one expected) as noted in Remark 2.7.

We introduce the concept of a uniformly locally finite cover of a manifold of bounded geometry. This is a natural extension of a locally finite cover. Uniformity means that we require a global bound K on the number of sets in the cover that intersect any small open ball.

Lemma 2.16 **(Uniformly locally finite cover)** *Let (M, g) be a Riemannian manifold of bounded geometry.*

Then for $\delta_2 > 0$ small enough and any $0 < \delta_1 \le \delta_2$, M has a countable cover $\{B(x_i; \delta_1)\}_{i \ge 1}$ such that

1. $\forall\, i \ne j : d(x_i, x_j) \ge \delta_1$;
2. *there exists an explicit global bound K such that for each $x \in M$ the ball $B(x; \delta_2)$ intersects at most K of the $B(x_i; \delta_2)$.*

Note that the second result implies both that the cover is locally finite with fixed neighborhood size, and that each set in the cover overlaps with at most K others, cf. Lebesgue covering dimension.

Proof Using Proposition 2.5, choose $\delta > 0$ such that Euclidean distance in normal coordinates on each $B(x; \delta)$ is $C = 2$ equivalent to the metric distance and set $\delta_2 \le \delta/3$.

Let $\{M_k\}_{k\in\mathbb{N}}$ be a compact exhaustion of M. Cover M_k with a sequence of balls $B(x_i;\delta_1)$, where $d(x_i,x_j)\geq\delta_1$. This sequence is finite, because an infinite sequence $\{x_i\}_{i\geq 0}$ must have an accumulation point in M_k, which contradicts $d(x_i,x_j)\geq\delta_1$. Choosing the first x_i's in M_{k+1} to coincide with those of M_k, it follows that the union of all balls $B(x_i,\delta_1)$ is a countable cover of M such that $\forall\, i\neq j: d(x_i,x_j)\geq\delta_1$.

Let $x\in M$ arbitrary. Any ball $B(x_i;\delta_2)$ that intersects $B(x;\delta_2)$ must be completely contained in $B(x;3\,\delta_2)$. Each of these balls has an exclusive subset $B(x_i;\delta_1/2)$, so in normal coordinates around x, each has an exclusive volume of at least $\mathrm{Vol}\big(B(0;\delta_1/(2C))\big)$, while $B(x;\delta)$ has volume of at most $\mathrm{Vol}\big(B(0;C\,3\,\delta_2)\big)$. With $n=\dim(M)$, this leads to the explicit upper bound

$$K\leq\frac{(3\,C\,\delta_2)^n}{(\delta_1/(2\,C))^n}=\Big(24\,\frac{\delta_2}{\delta_1}\Big)^n. \tag{2.10}$$

Thus, only finitely many can intersect $B(x;\delta_2)$. These estimates are uniform and do not depend on $x\in M$ so the bound K is global. □

Lemma 2.17 (Uniform partition of unity) *Let M be a manifold with a uniformly locally finite cover with $\delta_1<\delta_2$ and δ_2 sufficiently small, as per Lemma 2.16.*

Then there exists a partition of unity by functions $\chi_\bullet\in C^k_{b,u}(B(x_i;\delta_2);[0,1])$ subordinate to this cover.

We shall also apply this lemma to submanifolds which have a uniformly locally finite cover due to Corollary 2.26 on page 54.

Proof Let δ_2 be small enough that by Lemma 2.6 coordinate transition maps are $C^k_{b,u}$. Define a standard radially symmetric smooth bump function $\varphi\in C^\infty(\mathbb{R}^n;[0,1])$ that is identically one on $B(0;\delta_1)$ and has compact support in $B(0;\delta_2)$, hence $\varphi\in C^k_{b,u}$. We set $\varphi_i=\varphi\circ\exp_{x_i}^{-1}$ by isometric identification $\mathrm{T}_{x_i}M\cong\mathbb{R}^n$ and zero outside $B(x_i;\delta_2)$. We have $\varphi_\bullet\in C^k_{b,u}$ in any coordinate patch. Define in the usual way

$$\chi_i=\varphi_i\Big/\sum_{n\geq 1}\varphi_n. \tag{2.11}$$

The sum is finite as at most K of the $B(x_n;\delta_2)$ overlap any $B(x_i;\delta_2)$. The balls $B(x_i;\delta_1)$ already cover M, so the denominator is at least one, from which it follows that $\chi_\bullet\in C^k_{b,u}$. □

Corollary 2.18 *Similar to a uniform partition of unity, we can construct a partition by functions $\chi_\bullet\in C^k_{b,u}(B(x_i;\delta_2);[0,1])$ whose squares sum to one.*

In the proof of Lemma 2.17 we simply replace (2.11) by

$$\chi_i=\varphi_i\Big/\sqrt{\sum_{n\geq 1}\varphi_n^2}. \tag{2.12}$$

2.2 Curvature and Holonomy

To prove smoothness of the persistent manifold in Sect. 3.7, we shall want to estimate the holonomy along closed loops to be close to the identity, that is, if c is a closed loop, then we want $\Pi(c) - \mathbb{1}$ to be small. To this end, we relate the holonomy to the curvature and finally obtain an estimate in terms of a global bound on the curvature and the area of a surface enclosed by c.

The result that curvature is the generator of holonomy dates back at least to Ambrose and Singer [AS53] who formulated this in differential form in the 1950's; they cite an even older statement (without proof) by Élie Cartan [Car26]. More recent work by Reckziegel and Wilhelmus [RW06] shows explicit integral formulas for this relation, formulated on fiber bundles, a context far more general than is required here. We shall present a formulation for Riemannian manifolds (M, g).

Let Π denote the parallel transport functional, which takes C^1 curves to orthogonal maps between the tangent spaces at their endpoints, see (F.3). If c is a closed loop, then $\Pi(c)$ is a linear endomorphism on $\mathrm{T}_{c(0)}M$ and we can measure $\|\Pi(c) - \mathbb{1}\|$. Our goal is to bound this quantity by the integral of the curvature form R over a surface with boundary precisely c. This result can be viewed as a generalization of Stokes' theorem where the curvature is the exterior derivative of the connection form ω, while the connection on the other hand generates parallel transport along the boundary of the surface A that the curvature is integrated over. Note though, that we actually have $R = \mathrm{d}\omega + \omega \wedge \omega$, so there is an additional term due to the noncommutativity of the connection form.

Let

$$\gamma : D = \left[0, \bar{t}\right] \times [0, \bar{s}] \to M : (t, s) \mapsto \gamma(t, s) \tag{2.13}$$

parametrize the surface $A = \gamma(D) \subset M$. The idea is that γ is the homotopy of a (closed) curve c. We shall only consider parallel transport along horizontal or vertical lines in D; let us denote by $\Pi_t^{s_2, s_1}$ parallel transport along $s \mapsto \gamma(s, t)$ with $s \in [s_1, s_2]$ and by Π_{t_2, t_1}^{s} parallel transport along $t \mapsto \gamma(s, t)$ with $t \in [t_1, t_2]$.

We shall calculate the holonomy along ∂A with respect to a chosen frame on the pullback bundle $\gamma^*(\mathrm{T}M)$. The final result will turn out to be independent of this choice, hence it is covariantly defined. Let f be an orthonormal frame on $\gamma^*(\mathrm{T}M)$, that is, $f_{t,s} : \mathrm{T}_{\gamma(t,s)}M \to \mathbb{R}^n$ is an isometry of inner product spaces. The Levi-Civita connection ∇ on M can be pulled back to the connection $\gamma^*(\nabla)$ on $\gamma^*(\mathrm{T}M)$ and it can be expressed in terms of the connection form $\omega \in \Omega^1\big(D; \mathrm{End}(\mathbb{R}^n)\big)$ with respect to the frame f. The curvature of $\gamma^*(\nabla)$ is equal to the curvature R of ∇ pulled back to D, so we have $\mathrm{d}\omega + \omega \wedge \omega = \gamma^*(R)_f$, where the subscript f indicates that everything is expressed with respect to the chosen frame. In the same notation, parallel transport along a curve $s \mapsto c(s)$ satisfies the linear, homogeneous differential equation[3]

[3] If the frame f is induced by local coordinates, then ω will precisely be given by the Christoffel symbols and we recover Eq. (F.4).

$$\frac{d}{ds} f\Pi(c|_0^s) = -\omega(\dot{c}(s)) \circ f\Pi(c|_0^s), \qquad f\Pi(c|_0^0) = \mathbb{1}_{\mathbb{R}^n}, \tag{2.14}$$

which has a unique solution $s \mapsto {}_f\Pi^{s,0} = {}_f\Pi(c|_0^s)$. This can be viewed as time-dependent flow in $\mathrm{End}(\mathbb{R}^n)$.

Let us define the parallel transport term

$$P(s) = \Pi_{\bar{t}}^{\bar{s},s} \circ \Pi_{0,\bar{t}}^{s} \circ \Pi_0^{s,0} : \mathrm{T}_{\gamma(0,0)}M \to \mathrm{T}_{\gamma(\bar{t},\bar{s})}M, \tag{2.15}$$

see Fig. 2.1. The holonomy defect can be expressed as

$$\mathbb{1} - \Pi(\partial A) = \mathbb{1} - \Pi(\partial D) = \mathbb{1} - P(\bar{s})^{-1} \circ P(0) = P(\bar{s})^{-1} \circ \big(P(\bar{s}) - P(0)\big),$$

where $\Pi(\partial D)$ is defined using the pullback connection. We use the fundamental theorem of calculus to write

$$P(\bar{s}) - P(0) = \int_0^{\bar{s}} \frac{\mathrm{d}P(s)}{\mathrm{d}s} \mathrm{d}s. \tag{2.16}$$

Expressing everything with respect to the frame f, we see that the first and last factor of $P(s)$ are easily differentiated using (2.14):

$$\frac{\mathrm{d}}{\mathrm{d}s} {}_f\Pi_0^{s,0} = -\omega(\frac{\partial}{\partial s}) \circ {}_f\Pi_0^{s,0} \quad \text{and} \quad \frac{\mathrm{d}}{\mathrm{d}s} {}_f\Pi_{\bar{t}}^{\bar{s},s} = {}_f\Pi_{\bar{t}}^{\bar{s},s} \circ \omega(\frac{\partial}{\partial s}). \tag{2.17}$$

The middle term ${}_f\Pi_{\bar{t},0}^{s}$ can be differentiated by viewing s as parameter in the differential Eq. (2.14). Variation of constants yields (see e.g. [DK00, App. B] for a proof of the differentiable dependence of a flow on parameters)

$$\begin{aligned} \frac{\mathrm{d}}{\mathrm{d}s} {}_f\Pi_{\bar{t},t}^{s} &= \int_0^{\bar{t}} {}_f\Pi_{\bar{t},t}^{s} \circ \frac{\mathrm{d}}{\mathrm{d}s}\big[-\omega(\frac{\partial}{\partial t})\big] \circ {}_f\Pi_{t,0}^{s}\, \mathrm{d}t \\ &= \int_0^{\bar{t}} -{}_f\Pi_{\bar{t},t}^{s} \circ \Big(\mathrm{d}\omega(\frac{\partial}{\partial s}, \frac{\partial}{\partial t}) + \frac{\mathrm{d}}{\mathrm{d}t}\big[\omega(\frac{\partial}{\partial s})\big] + \omega([\frac{\partial}{\partial s}, \frac{\partial}{\partial t}])\Big) \circ {}_f\Pi_{t,0}^{s}\, \mathrm{d}t \end{aligned}$$

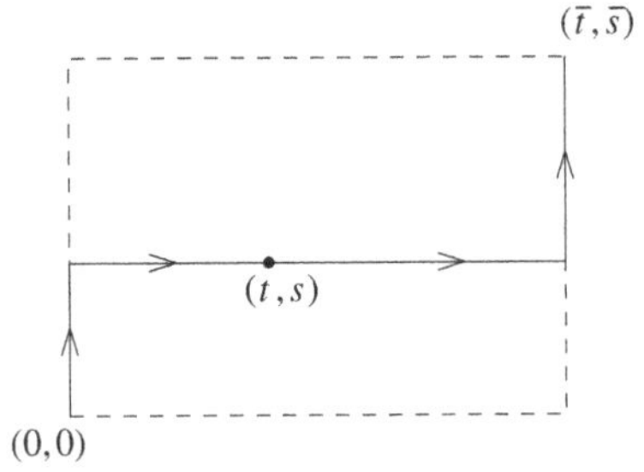

Fig. 2.1 The path of the parallel transport term $P(s)$ in D

using standard rules for exterior derivatives. Next we note that $\left[\frac{\partial}{\partial s}, \frac{\partial}{\partial t}\right] = 0$, and integrate by parts the term $\frac{\mathrm{d}}{\mathrm{d}t}\left[\omega(\frac{\partial}{\partial s})\right]$

$$
\begin{aligned}
&= \int_0^{\bar{t}} -{}_f\Pi^s_{\bar{t},t} \circ \Big(-\omega(\frac{\partial}{\partial t}) \circ \omega(\frac{\partial}{\partial s}) + \mathrm{d}\omega(\frac{\partial}{\partial s}, \frac{\partial}{\partial t}) + \omega(\frac{\partial}{\partial s}) \circ \omega(\frac{\partial}{\partial t}) \Big) \circ {}_f\Pi^s_{t,0}\, \mathrm{d}t \\
&\quad - \Big[{}_f\Pi^s_{\bar{t},t} \circ \omega(\frac{\partial}{\partial t}) \circ {}_f\Pi^s_{t,0} \Big]_{t=0}^{\bar{t}} \\
&= \int_0^{\bar{t}} {}_f\Pi^s_{\bar{t},t} \circ (\mathrm{d}\omega + \omega \wedge \omega)(\frac{\partial}{\partial t}, \frac{\partial}{\partial s}) \circ {}_f\Pi^s_{t,0}\, \mathrm{d}t \;\; - \omega(\frac{\partial}{\partial s}) \circ {}_f\Pi^s_{\bar{t},0} + {}_f\Pi^s_{\bar{t},0} \circ \omega(\frac{\partial}{\partial s}).
\end{aligned}
\tag{2.18}
$$

We see that this variation depends on the curvature form $\gamma^*(R)_f = \mathrm{d}\omega + \omega \wedge \omega$ along the path and two additional boundary terms. If we view γ as a homotopy of paths with homotopy parameter s and we keep the path endpoints $\gamma(0, s)$ and $\gamma(\bar{t}, s)$ fixed for all $s \in [0, \bar{s}]$, then these boundary terms vanish and the result (2.18) agrees with [RW06, Cor. 3].

Instead, we insert (2.17) and (2.18) into (2.16). Then these boundary terms cancel against the terms from (2.17) and we finally obtain

$$
\begin{aligned}
P(\bar{s})_f - P(0)_f &= \int_0^{\bar{s}} \int_0^{\bar{t}} {}_f\Pi^{\bar{s},s}_{\bar{t}} \circ {}_f\Pi^s_{\bar{t},t} \circ \gamma^*(R)_f(\frac{\partial}{\partial t}, \frac{\partial}{\partial s}) \circ {}_f\Pi^s_{t,0} \circ {}_f\Pi^{s,0}_0 \mathrm{d}t\mathrm{d}s \\
&= \Big(\int_D \Pi^{\bar{s},s}_{\bar{t}} \circ \Pi^s_{\bar{t},t} \circ \gamma^*(R) \circ \Pi^s_{t,0} \circ \Pi^{s,0}_0 \Big)_f .
\end{aligned}
\tag{2.19}
$$

The integrand on the last line is a two-form on D with values in $\mathcal{L}(\mathrm{T}_{\gamma(0,0)}M; \mathrm{T}_{\gamma(\bar{t},\bar{s})}M)$. This final expression is clearly independent of a choice of frame, so we have recovered an explicit integral formula relating holonomy along a null-homotopic loop to the curvature.

We conclude from (2.19) that if c is a closed, null-homotopic loop, and the curvature globally bounded, then $\|\Pi(c) - \mathbb{1}\|$ can be estimated by $\|R\|_{\sup}$ times the surface area of any null-homotopy γ of c. Note that we do not require γ to be an embedding; the integral is intrinsically defined on D by pullback. Furthermore, γ is required to be C^1 only. This follows from the fact that both sides of the equation are continuous with respect to γ in C^1-norm; alternatively, an explicit calculation requires that the mixed partial derivative $\frac{\partial^2 \gamma}{\partial s\, \partial t}$ is continuous to perform integration by parts. Both lead to the to the following result.

Lemma 2.19 (Exponential growth bound on holonomy) *Let (M, g) be a manifold of bounded geometry with normal coordinate radius δ that is M-small as in Definition 2.8. Fix $T > 0$ and $\rho > 0$ and let x_1, x_2 be two C^1 curves on M with derivatives bounded by N such that $d_\rho(x_1, x_2)\, e^{\rho T} \le \delta < r_{inj}(M)$. Denote by γ_t the unique shortest geodesic connecting $x_1(t)$ to $x_2(t)$ for any $t \in [0, T]$.*

If δ is sufficiently small, then the closed loop $\eta = x_2|_T^0 \circ \gamma_T \circ x_1|_0^T \circ \gamma_0^{-1}$ satisfies the holonomy bound

$$\|\Pi(\eta) - \mathbb{1}\| \le \tilde{C}\, \|R\|_0\, N\, d_\rho(x_1, x_2)\, \frac{e^{\rho\, T}}{\rho} \tag{2.20}$$

where $\tilde{C}$ depends on the geometry of M only.

Proof The two-parameter family $(s, t) \mapsto \gamma_t(s)$ defines a null-homotopy of the closed loop η. The map $s \mapsto \gamma_t(s)$ is defined through the exponential map as

$$\gamma_t : [0, 1] \to M : s \mapsto \exp_{x_1(t)}\big(s\, \exp_{x_1(t)}^{-1}(x_2(t))\big).$$

Since $\exp_x$ is a local diffeomorphism at least for $d(x_1(t), x_2(t)) < \delta\, e^{\rho t} < r_{\rm inj}(M)$, that depends smoothly on x, it follows that $(s, t) \mapsto \gamma_t(s)$ defines a homotopy between the curves x_1, x_2 restricted to the interval $[0, T]$. The map $\gamma_t(s)$ has continuous mixed derivatives with respect to s, t (even though the double derivative with respect to t does not exist since $x_1, x_2 \in C^1$ only), so integration by parts is allowed in (2.18).

We estimate the surface area mapped by $\gamma_t(s)$. We use shorthand notation $\xi = s\, \exp_{x_1(t)}^{-1}(x_2(t)) \in \mathrm{T}_{x_1(t)}M$ and denote by $\mathrm{D}_x \exp_x$ the derivative of the exponential map with respect to the base point parameter x. Then

$$\begin{aligned}
\frac{\mathrm{d}}{\mathrm{d}s}\gamma_t(s) =& \mathrm{D}\exp_{x_1(t)}(\xi) \cdot \exp_{x_1(t)}^{-1}(x_2(t)),\\
\frac{\mathrm{d}}{\mathrm{d}t}\gamma_t(s) =& \mathrm{D}_x \exp_{x_1(t)}(\xi) \cdot \dot{x}_1(t) + \mathrm{D}\exp_{x_1(t)}(\xi)\cdot\\
&\big[s\, \mathrm{D}_x(\exp_{x_1(t)}^{-1})(x_2(t)) \cdot \dot{x}_1(t) + s\, \mathrm{D}\exp_{x_1(t)}^{-1}(x_2(t)) \cdot \dot{x}_2(t)\big].
\end{aligned}$$

Since M has bounded geometry, $\mathrm{D}\exp_x$ and its inverse are bounded by Theorem 2.4, while $\mathrm{D}_x \exp_x$ and its inverse are bounded by Lemma 2.6, say by $C > 1$. This leads to estimates

$$\begin{aligned}
\left\|\frac{\mathrm{d}}{\mathrm{d}s}\gamma_t(s)\right\| &\le C\, d(x_1(t), x_2(t)),\\
\left\|\frac{\mathrm{d}}{\mathrm{d}t}\gamma_t(s)\right\| &\le C\, \|\dot{x}_1\|(t) + C\, s[C\, \|\dot{x}_1(t)\| + C\, \|\dot{x}_2(t)\|] \le 3\, C^2\, N,
\end{aligned}$$

so the holonomy bound satisfies

$$
\begin{aligned}
\|\Pi(\eta) - \mathbb{1}\| &\le \|R\|_0 \int_0^1 \int_0^T \|\frac{\mathrm{d}}{\mathrm{d}s}\gamma_t(s)\| \, \|\frac{\mathrm{d}}{\mathrm{d}t}\gamma_t(s)\| \mathrm{d}t\mathrm{d}s \\
&\le \|R\|_0 \int_0^T 3\,C^3\,N\,d_\rho(x_1, x_2)\,e^{\rho\,t}\mathrm{d}t \\
&\le 3\,C^3\,\|R\|_0\,N\,d_\rho(x_1, x_2)\,\frac{e^{\rho\,T}}{\rho}.
\end{aligned}
$$

□

Remark 2.20 It should be possible to obtain $\tilde{C} = 1$ if the curves x_i are generated by a flow Φ and we choose as homotopy $(s, t) \mapsto \Phi^t(\gamma(s))$, where γ is the geodesic connecting $x_1(0)$ and $x_2(0)$. In our applications, though, the curves x_1, x_2 need not be solutions to exactly the same flow, while the current result is sufficient for our purposes. ◇

2.3 Submanifolds and Tubular Neighborhoods

From this section on, we shall prove results that—although they may be of interest independently within bounded geometry—are building up towards the final section of this chapter, where we prove how to reduce Theorem 3.1 on persistence in general manifolds of bounded geometry to the setting of a trivial bundle. These results form the more technical part of this chapter and are not required elsewhere.

In the following, we assume that (Q, g) is an ambient manifold that has bounded geometry of large or infinite order and $M \in C^k$ will denote a submanifold of Q. Only a finite order $l > k$ of bounded geometry is required of (Q, g), but for simplicity we shall assume $l = \infty$. Recovering the explicit additional order $l-k$ would amount to tediously tracking the details throughout all the proofs; it should be sufficient if l is larger than k by some number between 2 and 10.

Let $\iota : M \to Q$ be a C^1 immersion. With abuse of notation we denote by $\mathrm{T}_x M = \mathrm{Im}(\mathrm{D}\iota(x))$ and $N_x = \mathrm{Im}(\mathrm{D}\iota(x))^\perp$ the tangent and normal spaces of M with respect to the immersion. Note that even if ι is not injective, the original point $x \in M$ uniquely selects the tangent and normal spaces in $\mathrm{T}_{\iota(x)} Q$.

Definition 2.21 (Uniformly immersed submanifold) *Let $\iota : M \to Q$ be a $C^{k\ge 1}$ immersion of M into the Riemannian manifold (Q, g) of bounded geometry. Denote by $M_{x,\delta}$ the image under ι of the connected component of x in $\iota^{-1}\big(B(\iota(x); \delta) \cap \iota(M)\big)$. We define M to be a $C^k_{b,u}$ immersed submanifold when there exists a $\delta > 0$ such that for all $x \in M$, the connected component $M_{x,\delta}$ is represented in normal coordinates on $B(\iota(x); \delta) \subset Q$ by the graph of a function $h_x : \mathrm{T}_x M \to N_x$ and the family of functions $h_\bullet \in C^k_{b,u}(\mathrm{T}_\bullet M; N_\bullet)$ has uniform continuity and boundedness estimates independent of x. We define $C^{k\ge 1}_b$ immersions in a similar way.*

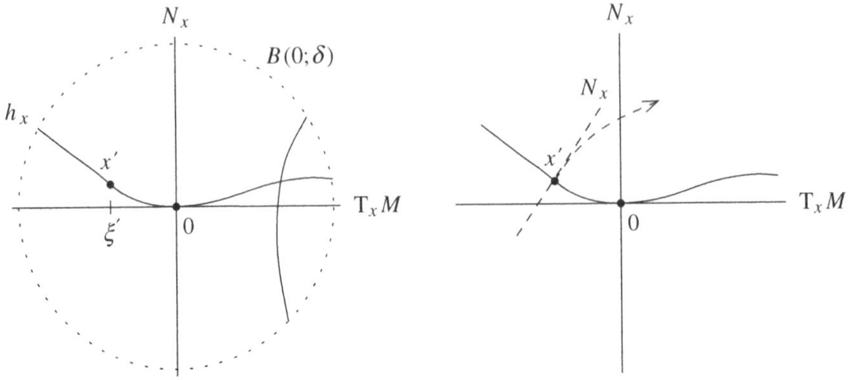

Fig. 2.2 An immersed submanifold represented by the graph of h_x in normal coordinates. In the left figure, another part of M intersects transversely on the *right*; the *right figure* contains an orbit of the geodesic flow along a normal vector at $x' \cong \iota(x')$

Remark 2.22 By taking the connected component $M_{x,\delta}$ in M, we allow for immersed submanifolds that intersect, or nearly intersect themselves. See Fig. 2.2 on the left: $M_{x,\delta}$ is described by the graph of h_x, while on the right side, a different part of M embeds into this same neighborhood $B(\iota(x);\delta)$. See Fig. 3.2 on page 86 for an example of a nearly self-intersecting submanifold. If we want to rule out such cases, we can assume that $M_{x,\delta}$ is the unique component of $M \cap B(\iota(x);\delta)$. This will turn M into an embedded submanifold, but more strongly, the nearly self-intersecting case is also ruled out. We will refer to this as a *uniformly embedded submanifold*. ◇

Remark 2.23 The sets $M_{x,\delta}$ play a similar role as 'plaques' in [HPS77, p. 72–73]. ◇

Remark 2.24 In case $k = 1$, boundedness is automatically implied by uniform continuity. This follows from the representation in normal coordinates. We have $\mathrm{D}h_x(0) = 0$, so by uniform continuity there exists a $\delta > 0$ such that $\|\mathrm{D}h_x(\xi)\| < \varepsilon = 1$ when $\|\xi\| < \delta$, hence $\mathrm{D}h_x$ is bounded. Put another way, there is no intrinsic measure for the 'size of the derivative or tangent' of a submanifold. ◇

Note that the function h_x is only defined on that part of the domain $B(0;\delta) \subset \mathrm{T}_x M$ where its graph is contained in $B(0;\delta) \subset \mathrm{T}_{\iota(x)} Q$, as can be seen in Fig. 2.2. In the splitting $\mathrm{T}_{\iota(x)} Q = \mathrm{T}_x M \oplus N_x$, we denote with p_1, p_2 orthogonal projections onto the $\mathrm{T}_x M$ and N_x subspaces, respectively.

From now on we shall continually assume that $M \in C_{b,u}^{k\geq 1}$ is a uniformly immersed submanifold of Q. We will often identify M with its image $\iota(M) \subset Q$, as well as identify points $x \in M$ with $\iota(x)$, keeping in mind the definition of $M_{x,\delta}$ to track local injectivity. Furthermore, denote by d_M the distance on M induced by the pulled back Riemannian metric $\iota^*(g)$. This distance function measures whether points are close when viewed along the domain of the immersion, disallowing 'shortcuts' through Q.

It also distinguishes different points with the same immersion image. Note that it is different from the distance d on Q pulled back to M. This we denote by $d_Q = \iota^*(d)$ but it is not a distance on M when ι is not injective. Still, we have the following local result, which will be useful for later estimates.

Lemma 2.25 (Local equivalence of distance) *Let $M \in C^1_{b,u}$ be a uniformly immersed submanifold of the bounded geometry manifold (Q, g). Then d_Q and d_M are locally equivalent in the following sense:*

1. $\forall x_1, x_2 \in M : d_Q(x_1, x_2) \leq d_M(x_1, x_2)$;
2. *for any $C' > 1$ there exists a $\delta > 0$ such that for all $d_M(x_1, x_2) < \delta$, we have the local converse $d_M(x_1, x_2) \leq C' d_Q(x_1, x_2)$.*

Proof The first assertion follows directly from the fact that any path in M induces a path of equal length in Q via the immersion ι.

For the second part, we first note that if δ is small enough and $d_M(x_1, x_2) < \delta$, then we must have $x_2 \in M_{x_1,\delta}$. If this would not be the case, then any path γ connecting x_1, x_2 through M cannot be contained in $M_{x_1,\delta}$. But this implies that the path runs out of $B(x_1; \delta)$, so its length is greater than δ. This contradicts the assumption that $d_M(x_1, x_2) < \delta$. Hence, x_2 can be represented as a point on the graph of h_{x_1} in $B(x_1; \delta)$.

Let $C > 1$, $\varepsilon > 0$ be constants to be fixed later and let δ be small enough such that the metric coefficients are bounded by C in normal coordinate charts, that Proposition 2.5 holds with C, and we have $\|h_\bullet\|_1 \leq \varepsilon$ as in Remark 2.24. We consider the normal coordinate chart on $B(x_1; \delta)$ and construct a path in M to find an upper bound for $d_M(x_1, x_2)$. Let $x_2 = (\xi, h_{x_1}(\xi))$ and define $\gamma(t) = (t\,\xi, h_{x_1}(t\,\xi))$ for $t \in [0, 1]$. We estimate the length of γ as

$$l(\gamma) \leq \int_0^1 \sqrt{\|g\|}\sqrt{1 + \|h_\bullet\|_1^2}\,\|\xi\|\mathrm{d}t \leq \sqrt{C}\sqrt{1 + \varepsilon^2}\,\|\xi\|,$$

while the Euclidean norm can be estimated by the distance in Q as

$$\|\xi\| \leq \|(\xi, h_{x_1}(\xi))\| \leq C\,d(x_1, x_2).$$

We conclude that

$$d_M(x_1, x_2) \leq l(\gamma) \leq C^{3/2}\sqrt{1 + \varepsilon^2}\,d_Q(x_1, x_2)$$

and for any $C' > 1$ we can find $C > 1$, $\varepsilon > 0$ such that $C^{3/2}\sqrt{1 + \varepsilon^2} < C'$. □

A uniform submanifold of a bounded geometry manifold can be shown to possess a uniformly locally finite cover as a corollary of Lemma 2.16, without the need to show that the submanifold itself has bounded geometry. As a consequence, it also has (square-sum) partitions of unity.

Corollary 2.26 (Uniform cover of a submanifold) *Let $M \in C^1_{b,u}$ be a uniformly immersed submanifold of the bounded geometry manifold (Q, g).*

Then for $\delta_2 > 0$ small enough and any $\delta_1 \in (0, \delta_2]$, M has a uniformly locally finite cover by balls of radius δ_2 in terms of the distance d_Q, such that the balls of radius δ_1 already cover M. That is, there exist $\{x_i\}_{i\geq 1}$ such that $\bigcup_{i\geq 1} M_{x_i,\delta_1}$ covers M with a uniform bound K on the maximum number of sets M_{x_i,δ_2} covering any set M_{x,δ_2} with $x \in M$.

Proof The proof follows the ideas of Lemma 2.16. As an additional requirement, let $\delta > 0$ be sufficiently small such that each $M_{x,\delta}$ is represented in normal coordinates by the graph of h_x. Under this assumption, the open sets $M_{x,\delta}$ are induced by d_Q and correspond to the connected component of x of the pre-image of $B(\iota(x); \delta)$. Consequently, we can locally push the argument to $\iota(M) \subset Q$ to conclude that there is an upper bound K on the number of sets M_{x_i,δ_2} that intersect any set M_{x,δ_2}. □

Even though we do not require submanifolds to have bounded geometry for the results in this section, the lemma below will be needed in the final reduction to a trivial bundle. The essential idea of the proof is to use Gauß' second fundamental form to relate curvature of the submanifold to second derivatives of its immersion map.

Lemma 2.27 (Submanifold of bounded geometry) *Let $M \in C^{k\geq 2}_b$ be a uniformly immersed submanifold of the bounded geometry manifold (Q, g). Then $(M, \iota^*(g))$ is a Riemannian manifold with bounded geometry of order $k - 2$.*

Remark 2.28 We lose two orders of smoothness in the bounded geometry definition. This is due to bounded geometry being defined in terms of the curvature, which depends on second order derivatives of the metric, and in this case also on second order derivatives of the embedding through Gauß' second fundamental form. ◇

Proof Let δ be sufficiently small such that for each $x \in M$ we have the representation $M_{x,\delta} = \mathrm{Graph}(h_x)$ with $\|\mathrm{D}h_x\| \leq 1$.

The Riemann curvature tensor R^M of M can be expressed as a sum of the curvature R on Q and the second fundamental form of the (local) embedding, see e.g. [Jos08, Thm. 3.6.2]:

$$\begin{aligned} g(R^M(X, Y)\, Z, W) &= g(R(X, Y)\, Z, W) + g(S(Y, Z), S(X, W)) \\ &\quad - g(S(Y, W), S(X, Z)), \end{aligned} \tag{2.21}$$

where

$$S : \mathrm{T}M \times \mathrm{T}M \to N : X, Y \mapsto (\nabla_X Y)^\perp \tag{2.22}$$

is the second fundamental form, and it is indeed pointwise defined. In normal coordinates we find

$$S_x(X, Y) = \mathrm{D}^2 h_x(0)(X, Y). \tag{2.23}$$

Since $h \in C_b^k$ and $g,\ g^{-1} \in C_b^k$ as well, it follows that $S \in C_b^{k-2}$ and by (2.21) then that $R^M \in C_b^{k-2}$, so condition (B_{k-2}) of Definition 2.1 is satisfied.

Condition (I) on the injectivity radius follows from an implicit function argument applied to the geodesic flow using Theorem A.6. We consider local coordinates around $x \in M$ by projecting the representation $M \cap B(x;\delta)$ onto $\mathrm{T}_x M$ in normal coordinates in Q. That is, we have the coordinate chart map

$$\kappa_x : B(0;\delta/2) \subset \mathrm{T}_x M \to M : \xi \mapsto \exp_x(\xi, h_x(\xi))$$

and the corresponding embedding into normal coordinates $\mathrm{T}_x M \hookrightarrow \mathrm{T}_x Q : \xi \mapsto (\xi, h_x(\xi))$ of Q. We calculate explicit estimates for the exponential map $\exp_x^M$ using Christoffel symbols of the connection ∇^M on M in the coordinates in chart κ_x.

Let X, Y be vector fields on M. Their representation in κ_x is mapped to normal coordinates $B(x;\delta)$ on Q as

$$X(\xi) \mapsto \tilde{X}(\xi) = \big(\mathbb{1}, \mathrm{D}h_x(\xi)\big)^T \cdot X(\xi).$$

Hence, from the covariant derivative on M in normal coordinates $B(x;\delta) \subset Q$ we can recover the Christoffel symbols in local coordinates κ_x as

$$\nabla_X^M Y = p_1 \circ \Big[X^i(\xi)\frac{\partial}{\partial \xi^i} Y(\xi) + \Gamma(\xi, h_x(\xi))\big(\tilde{X}(\xi), \tilde{Y}(\xi)\big)\Big],$$

where the first term has reduced to derivatives with respect to $\xi \in \mathrm{T}_x M \subset \mathrm{T}_x Q$ only, and $\Gamma : B(x;\delta) \to \mathcal{L}^2(\mathrm{T}_x Q; \mathrm{T}_x Q)$ are the Christoffel symbols in normal coordinates at $x \in Q$. Thus, the Christoffel symbols

$$\Gamma^M(\xi)(X,Y) = p_1 \circ \Gamma(\xi, h_x(\xi))\big(\tilde{X}, \tilde{Y}\big) \tag{2.24}$$

of M in κ_x coordinates are uniformly bounded on sufficiently small balls $B(0;\delta') \subset \mathrm{T}_x M$. The Euclidean geodesic flow at time one defines the (trivial) Euclidean exponential map, which is an isomorphic diffeomorphism (with infinite injectivity radius actually). Since we study a small perturbation of this flow in local coordinates, given by the additional term (2.24), and the perturbation is at least C_b^{k-1} and C^1 small, the perturbed geodesic flow of M can be made close enough that $\exp_x^M$ is still a diffeomorphism on $B(0;\delta')$ for some $\delta' > 0$. Hence, $r_{\mathrm{inj}}(x) \geq \delta'$, but these estimates depend only on the perturbation size, so they hold uniformly for all $x \in M$. □

To obtain the final result of this section, the tubular neighborhood theorem 2.33, we first need to work out some details on local coordinates. If M is a submanifold of Q, it is natural to consider a specific splitting on the normal coordinates at points $x \in M$, namely $\mathrm{T}_x Q = \mathrm{T}_x M \oplus N_x$, where N is the normal bundle over M. We shall require bounds, not just on coordinate transformations, but more specifically bounds on how well this splitting is preserved. The lemmas are formulated in a more general context of splittings of tangent spaces at any two nearby points, while the results for coordinates along M follow as an easy corollary.

Lemma 2.29 **(Coordinate transformations of splittings)** *Let (Q, g) be a smooth Riemannian manifold of bounded geometry, let C be sufficiently large and let $\delta, \zeta > 0$ be sufficiently small. Let $x_1, x_2 \in Q$ and let $\mathrm{T}_{x_i} Q = H_i \oplus V_i$, $i = 1, 2$ be splittings along 'horizontal' and 'vertical' perpendicular subspaces with $\dim(H_1) = \dim(H_2)$. Assume that $d(x_1, x_2) < \delta$ and that, for $i \neq j$, H_i is represented in tangent normal coordinates at x_j by the graph of $L_i \in \mathcal{L}(H_j; V_j)$ with $\|L_i\| \leq \zeta$.*

Then the coordinate transformation $\varphi_{2,1}$ in Lemma 2.6 is of the form

$$\varphi_{2,1} = O_H \oplus O_V + \tilde{\varphi}_{2,1} \quad \textit{with} \quad \|\tilde{\varphi}_{2,1}\|_k \leq C\big(\zeta + d(x_1, x_2)\big), \tag{2.25}$$

where O_H, O_V are orthogonal transformations between the H_i and V_i with $i = 1, 2$, respectively.

We first prove the following result and use it to prove Lemma 2.29.

Lemma 2.30 **(Approximation of orthogonal maps)** *Let V be a finite-dimensional inner product space and define the map*

$$f : \mathcal{L}(V) \to Sym(V) : A \mapsto A^T A - \mathbb{1}. \tag{2.26}$$

There exists an $\varepsilon > 0$ and a tubular neighborhood $B(O(V); \eta) \subset GL(V)$ with fiber projection π, such that on $\{A \in GL(V) \mid \|f(A)\| < \varepsilon, \|A\| \leq 2\}$, the map $\varphi : A \mapsto (\pi(A), f(A))$ is a smooth diffeomorphism. As a direct corollary, if $\|f(A)\| < \varepsilon$ and $\|A\| \leq 2$ then $U = \pi(A) \in O(V)$ is an orthogonal approximation of A in the sense that $\|U - A\| \leq \|f(A)\|$.

Proof The map f is smooth and invariant under the left action of the orthogonal maps $O(V)$, while $O(V) = \ker(f)$. Since $GL(V)$ is a Lie group, we have the canonical trivialization $\mathrm{T}GL(V) = GL(V) \times gl(V)$ by left multiplication. The similar trivialization $\mathrm{T}O(V) = O(V) \times o(V)$ can be viewed as a subbundle of

$$O(V) \times o(V) \oplus \mathrm{Sym}(V) = \mathrm{T}GL(V)|_{O(V)},$$

where $o(V)$ is identified with the skew-symmetric linear maps. We restrict the exponential map $\exp : \mathrm{T}GL(V) \to GL(V)$ to $O(V) \times \mathrm{Sym}(V)$. At $\mathbb{1} \in O(V)$ this restriction has bijective derivative, hence it is a local diffeomorphism. Since exp is $O(V)$-invariant, it defines a diffeomorphism onto a tubular neighborhood $B(O(V); \eta) \subset GL(V)$ of $O(V)$ of size $\eta > 0$ and a corresponding smooth fiber projection map $\pi : B(O(V); \eta) \to O(V)$.

Now $\mathrm{D}f(\mathbb{1}) : a \mapsto a^T + a$ has image precisely $\mathrm{Sym}(V)$. Thus, if we restrict f to the fiber over $\mathbb{1} \in O(V)$ in the tubular neighborhood, then $\mathrm{D}f(\mathbb{1})|_{\mathrm{Sym}(V)} = 2$ and f is a diffeomorphism with $\|\mathrm{D}f^{-1}\| \leq 1$ in some neighborhood of $0 \in \pi^{-1}(\mathbb{1})$; if necessary, we reduce $\eta > 0$ for $\|\mathrm{D}f^{-1}\| \leq 1$ to hold on $B(O(V); \eta) \cap$

$\pi^{-1}(\mathbb{1})$. By $O(V)$ invariance of f, this holds globally on all (fibers) of the tubular neighborhood. Since, $\mathrm{D}\pi$ and $\mathrm{D}f$ have complementary image at $O(V)$, $\varphi = (\pi, f)$ is a diffeomorphism on $B(O(V); \eta)$.

The set $\overline{B(0; 2) \setminus B(O(V); \eta)} \subset \mathcal{L}(V)$ is compact, so $\|f(\cdot)\|$ attains its nonzero minimum on it. Let ε be smaller than this minimum. Then, if $\|f(A)\| < \varepsilon$, we must have $A \in B(O(V); \eta)$ and hence $A = \exp(U, a)$ for a unique $(U, a) \in O(V) \times \mathrm{Sym}(V)$. By $O(V)$-invariance, we can assume w.l.o.g. that $U = \mathbb{1}$ and use the mean value theorem to estimate

$$\|A - \mathbb{1}\| \le \left\|\mathrm{D}f^{-1}\right\| \|f(A) - f(\mathbb{1})\| \le \|f(A)\| < \varepsilon.$$

In other words, when A is sufficiently close to being orthogonal, measured according to f, then it is close to an orthogonal map U in operator norm. □

Proof of lemma 2.29 Extending the results of Lemma 2.6, let

$$O = \Pi(\gamma_{2,1}) : \mathrm{T}_{x_1} Q \to \mathrm{T}_{x_2} Q$$

denote the orthogonal linear map induced by parallel transport. We decompose $\varphi_{2,1} = O + \hat{\varphi}_{2,1}$, where $\hat{\varphi}_{2,1}$ can be made arbitrarily small. Moreover, we write

$$O = \begin{pmatrix} A & B \\ C & D \end{pmatrix} \in \mathcal{L}(H_1 \oplus V_1; H_2 \oplus V_2),$$

with the idea that B, C should be small and A, D should approximate orthogonal maps O_H, O_V, respectively. Orthogonality of O implies

$$\mathbb{1} = O^T O = \begin{pmatrix} A^T & C^T \\ B^T & D^T \end{pmatrix} \cdot \begin{pmatrix} A & B \\ C & D \end{pmatrix} = \begin{pmatrix} A^T A + C^T C & A^T B + C^T D \\ B^T A + D^T C & B^T B + D^T D \end{pmatrix}.$$

For the operator norm we have $\|A\|, \|B\|, \|C\|, \|D\| \le \|O\| = 1$, so if we assume for the moment that B, C can be made sufficiently small, then, by writing $A^T A - \mathbb{1} = -C^T C$ and $D^T D - \mathbb{1} = -B^T B$, Lemma 2.30 implies that we can find O_H, O_V such that

$$\begin{aligned} \|O - O_H \oplus O_V\| &\le \left\|O - \begin{pmatrix} A & 0 \\ 0 & D \end{pmatrix}\right\| + \left\|\begin{pmatrix} A & 0 \\ 0 & D \end{pmatrix} - \begin{pmatrix} O_H & 0 \\ 0 & O_V \end{pmatrix}\right\| \\ &\le \|B\| + \|C\| + \|A - O_H\| + \|D - O_V\| \\ &\le \|B\| + \|C\| + \left\|C^T C\right\| + \left\|B^T B\right\|. \end{aligned} \tag{2.27}$$

In normal coordinates around x_2 we have $H_1 = \mathrm{Graph}(L)$, so $C : H_1 \to V_2$ is represented by L in these coordinates. The metric g is close to the identity in these coordinates, so $C \cong L$ can be assumed bounded by $4\zeta \le 1$, as measured in the metric on Q. The same argument can be made for B^T by considering $\varphi_{1,2} = \varphi_{2,1}^{-1}$, since

$$O^{-1} = O^T = \begin{pmatrix} A^T & C^T \\ B^T & D^T \end{pmatrix}.$$

We conclude that both $\|B\|, \|C\| \leq 4\zeta$ when δ is chosen small, hence O can be approximated by $O_H \oplus O_V$, and the error from (2.27) can be absorbed into $\tilde{\varphi}_{2,1}$:

$$\tilde{\varphi}_{2,1} = \hat{\varphi}_{2,1} + (O - O_H \oplus O_V).$$

The errors introduced in $\tilde{\varphi}_{2,1}$ from lemmas 2.6 and 2.30 are Lipschitz small in terms of $d(x_1, x_2)$ and ζ, respectively, so these add up to the estimate in (2.25). □

Corollary 2.27 *Let $M \in C^{k\geq 1}_{b,u}$ be a uniformly immersed submanifold of a smooth Riemannian manifold (Q, g) of bounded geometry. Let $x_1, x_2 \in M$ and let $\mathrm{T}_{x_i} Q = \mathrm{T}_{x_i} M \oplus N_{x_i}$, $i = 1, 2$, be the respective splittings in horizontal and vertical directions. Then the results of Lemma 2.29 hold for $d_M(x_1, x_2) < \delta$. If moreover $M \in C^2_b$, then we have a Lipschitz estimate $\|\mathrm{D}\tilde{\varphi}_{2,1}(0)\| \leq C\, d(x_1, x_2)$.*

Proof This follows immediately from the local representation $M_{x_2,\delta} = \mathrm{Graph}(h_2)$ as $\mathrm{T}_{x_1} M$ is represented in tangent normal coordinates at x_2 by $L = \mathrm{D}h_2(\xi)$, where $x_1 = (\xi, h_2(\xi))$. And $\mathrm{D}h_2(\xi)$ becomes small when δ is small. The same holds with x_1, x_2 interchanged.

If $M \in C^2_b$, then we can estimate $\|\mathrm{D}h_\bullet(\xi)\| \leq \|\mathrm{D}^2 h_\bullet\| \, \|\xi\| \leq C\, d(x_1, x_2)$. Hence, the Lipschitz result in Lemma 2.29 transforms into a Lipschitz estimate in $d(x_1, x_2)$ only. □

Below we define when a mapping is approximately isometric, see for example also [Att94, p 505]. The Lyapunov exponents of a dynamical system are preserved under these quasi-isometries since the exponential growth dominates any bounded factors when measuring sizes. This property is required when we transfer a noncompact normally hyperbolic system to a different space and want normal hyperbolicity to be preserved.

Definition 2.30 (quasi-isometry) *Let M, N be manifolds with distance metrics d_M, d_N and let $\varphi : M \to N$ be a diffeomorphism. we call φ a C-quasi-isometry with $C > 1$, if*

$$\forall x, y \in M : C^{-1}\, d_M(x, y) \leq d_N(\varphi(x), \varphi(y)) \leq C\, d_M(x, y). \tag{2.28}$$

We simply call φ a quasi-isometry if there exists an unspecified $C > 1$.

We conclude this section with a version of the tubular neighborhood theorem that is appropriate in the bounded geometry setting.

Theorem 2.31 (Uniform tubular neighborhood) *Let $M \in C^{k\geq 2}_b$ be a uniformly immersed submanifold of the bounded geometry manifold (Q, g). Then for $\eta > 0$ sufficiently small (but depending explicitly on M and Q), the η-sized tubular* *neighborhood* $B(M;\eta) = \{y \in Q \mid d(y, M) \leq \eta\}$ *can be represented on the* η*-sized normal bundle* $N_{\leq\eta}$ *of* M *by a diffeomorphism* φ*, locally on each* $N_{\leq\eta}|_{M_{x,\delta}}$ *and we have* $\varphi, \varphi^{-1} \in C^{k-1}_{b,u}$ *(hence* φ *is a quasi-isometry).*

When moreover M *is uniformly embedded, i.e.* $M_{x,\delta} = M \cap B(x;\delta)$ *for each* $x \in M$ *as in Remark 2.22, then* φ *is a global diffeomorphism.*

In case M is compact, the standard proof uses the fact that the exponential map has bijective differential at the zero section, and then by compactness it must be a diffeomorphism on a uniform neighborhood $N_{\leq\eta}$ of the zero section. Here, to get a uniform neighborhood $N_{\leq\eta}$ on which $\varphi = \exp|_{N_{\leq\eta}}$ is a diffeomorphism, we require bounds on second order derivatives (that is curvature, cf. Lemma 2.27) of M so that it has curvature radius r bounded from below, hence cut locus points can only occur at least at distance r away from M, making φ injective for $\eta < r$. See Fig. 2.2 for a representation of a submanifold M in normal coordinates around x and a ray of the normal bundle at a nearby point x'.

Note that for an immersed submanifold, we define the normal bundle as

$$N = \big\{(x, \nu) \in M \times \mathrm{T}Q | \iota(x) = \pi(\nu),\ \nu \perp \mathrm{Im}\big(\mathrm{D}\iota(x)\big)\big\}. \tag{2.29}$$

This can again be viewed as immersed into $\mathrm{T}Q$.

Proof We set $\varphi = \exp|_N$ and in the following we will implicitly apply Theorem 2.4 and Proposition 2.5 to choose $0 < \delta \leq 1$ small enough such that the metric g up to its second order derivatives is bounded, as well as that the Christoffel symbols are bounded. Also, we choose δ such that M is uniformly locally representable by graphs according to Definition 2.21. We will in sequence prove local and global injectivity, surjectivity of φ and finally that $\varphi, \varphi^{-1} \in C^{k-1}_{b,u}$.

We claim that for some $\eta > 0$, φ is locally injective on $N_{\leq\eta}$, the normal bundle restricted to size η. Let $\nu, \nu' \in N$ be such that $\varphi(\nu) = \varphi(\nu')$ and denote by $x = \pi(\nu)$, $x' = \pi(\nu')$ their base points in M. We consider normal coordinates at x, hence we have $\nu = (0, \sigma_2)$ for some $\sigma_2 \in N_x$, while ν' is given by $(\sigma_1', \sigma_2') \in \mathrm{T}_x M \oplus N_x$. From Corollary 2.31 it follows for small δ that ν' is nearly mapped onto N_x in normal coordinates at x. Since $M \in C^2_b$, the deviation from mapping onto N_x is Lipschitz small in $d(x', x) \cong \|\xi\|$, so we have

$$\|\sigma_1'\| \leq C\, \|\xi\|\, \|\sigma_2'\|. \tag{2.30}$$

Now, $\varphi(\nu') = \varphi(\nu)$ can only hold if the respective horizontal coordinates along $\mathrm{T}_x M$ are equal. By definition of normal coordinates around x, we have $\exp(\nu) = (0, \sigma_2)$. Therefore it is sufficient to prove that some $\eta > 0$ exists as a lower bound for

$$\big\{\|\nu'\| \in \mathbb{R} \mid \varphi(\nu')_1 = 0, \pi(\nu') \neq x\big\}.$$

We view the exponential map as the time-one geodesic flow, which is given in local coordinates by (2.5). The geodesic flow along $\nu' = (\sigma_1', \sigma_2')$ starting at $(\xi', h(\xi'))$ is a small perturbation of the flow along $(0, \sigma_2')$ starting at $(0, 0)$. The latter has solution curve $t \mapsto (0, t\,\sigma_2') \in \mathrm{T}_x M \oplus N_x$.

By Theorem 2.4 we arrange for $\|g - \mathbb{1}\|, \|\Gamma\| \le 2$ and $\|\mathrm{D}\Gamma\| \le C$ in local coordinates. We have estimates

$$\begin{aligned} \big\|\Gamma(x'(t)) - \Gamma(x(t))\big\| &\le \|\mathrm{D}\Gamma\|\, \big\|x'(t) - x(t)\big\| \le C\, \big\|x'(t) - x(t)\big\|, \\ \big\|\sigma'(t)\big\| &\le \sqrt{g(\sigma'(t), \sigma'(t))} = \sqrt{g(\sigma'(0), \sigma'(0))} \le \sqrt{2}\, \big\|\sigma'(0)\big\|. \end{aligned}$$

With these, we obtain the Gronwall-like estimates

$$\begin{aligned} \frac{\mathrm{d}}{\mathrm{d}t}\big\|x'(t) - (0, t\,\sigma_2')\big\| &\le \big\|\sigma'\big\|(t) - (0, \sigma_2'), \\ \frac{\mathrm{d}}{\mathrm{d}t}\big\|\sigma'(t) - (0, \sigma_2')\big\| &\le \|\mathrm{D}\Gamma\|\, \big\|x'(t) - (0, t\,\sigma_2')\big\|\, \big\|\sigma'(t)\big\|^2 \\ &\quad + \|\Gamma\|\, \big(\big\|\sigma'(t)\big\| + \big\|(0, \sigma_2')\big\|\big)\, \big\|\sigma'(t) - (0, \sigma_2')\big\| \\ &\le \big(4\,C\,\eta^2 + 4\sqrt{2}\,\eta\big)\, \big\|\sigma'(t) - (0, \sigma_2')\big\|, \end{aligned}$$

for which Gronwall's inequality yields

$$\big\|\sigma'(t) - (0, \sigma_2')\big\| \le \big\|(\sigma_1'(0), 0)\big\|\, e^{\tilde{C}\,\eta\,t}.$$

Now, if η is chosen sufficiently small, then using (2.30), we have for all $0 \le t \le 1$ that

$$\big\|x_1'(t) - x_1(0)\big\| \le \int_0^t C\, \big\|\xi'\big\|\, \eta\, e^{\tilde{C}\,\eta\,t} \mathrm{d}\tau \le \frac{1}{2}\, \big\|\xi'\big\|.$$

This shows that there exists an explicit $\eta > 0$ such that φ is injective on $N_{\le\eta}$ restricted to a neighborhood $M_{x,\delta}$, and by construction this η is uniform over M. Later modifications to choose η smaller will only depend on the global geometry of Q, but not on any details of M.

If moreover $M_{x,\delta} = M \cap B(x; \delta)$ is the unique connected component of M in each normal coordinate chart, then φ is injective globally on $N_{\le\eta}$. This follows easily by taking $\eta < \frac{\delta}{2}$. Then, any ν' and ν that have the same image, must have base points x', x separated by a distance less than δ, as $\varphi = \exp$ will only map onto points at most η away from the base point. Therefore, x' must lie in $B(x; \delta)$ and in M, hence on $\mathrm{Graph}(h) = M_{x,\delta}$. This case was already treated.

Finally, we will show that φ is surjective onto $B(M; \eta)$ when $\eta < r_{\mathrm{inj}}(Q)$. Take $y \in B(M; \eta)$, then $M \cap B(y; r_{\mathrm{inj}}(Q))$ contains a nonempty compact set, so there exists an $x \in M$ such that $d(y, M) = d(y, x)$. This distance must be realized by a (unique) geodesic γ. We will derive a contradiction if $\gamma'(0) \notin N_x$, by showing that then the minimum distance is not attained at x. Let $(\xi_1, \xi_2) \in \mathrm{T}_x M \oplus N_x$ be the normalized tangent vector of $\gamma'(0)$. By assumption we have $\xi_1 \ne 0$, so $\|\xi_2\| < 1$. We parametrize

$$\gamma : [0, d(y, x)] \to \mathrm{T}_x Q : t \mapsto t\,(\xi_1, \xi_2)$$

by arc length in normal coordinates, thus $d(x, \gamma(t)) = t$. Consider the Euclidean distance in normal coordinates at x of $\gamma(t)$ to its vertical projection onto $M = \mathrm{Graph}(h)$. This shows that

$$d_E(\gamma(t), M) \leq \|t\,\xi_2 - h(t\,\xi_1)\| \leq t\,\|\xi_2\| + o(t\,\|\xi_1\|)$$

as $\mathrm{D}h(0) = 0$ and $h \in C_b^2$. The Euclidean distance is C-equivalent to the g-induced distance, so we have

$$\lim_{t \downarrow 0} \frac{d(\gamma(t), M)}{d(x, \gamma(t))} \leq \lim_{t \downarrow 0} C\,\|\xi_2\| + o(t)/t = C\,\|\xi_2\|.$$

By assumption $\|\xi_2\| < 1$, so we can restrict to a small enough neighborhood $B(x; \delta)$ such that $1 < C < \|\xi_2\|^{-1}$ and conclude that $d(\gamma(t), M) < d(x, \gamma(t))$ for some $t > 0$, which shows that a shorter (broken) geodesic from y to M exists. This completes the contradiction and proves that φ is surjective.

Finally, $\varphi \in C_{b,u}^{k-1}$ follows directly from the fact that it is the restriction of the exponential map to $N_{\leq\eta} \in C^{k-1}$ and in induced normal coordinate charts, we have $\exp \in C_{b,u}^{k-1}$. For φ^{-1} we use a formula and arguments similar to (A.2), showing that if $\mathrm{D}\varphi^{-1}$ is uniformly bounded, then $\varphi^{-1} \in C_{b,u}^{k-1}$ holds as well. Now $\mathrm{D}\varphi = \mathbb{1}$ in induced normal coordinates at M, so by uniform continuity, there exists some $\eta > 0$ such that $\mathrm{D}\varphi$ stays away from non-invertibility on $N_{\leq\eta}$, hence $\mathrm{D}\varphi^{-1}$ stays bounded. This automatically implies that φ is a quasi-isometry with $C = \max(\|\mathrm{D}\varphi\|, \|\mathrm{D}\varphi^{-1}\|)$. □

2.4 Smoothing of Submanifolds

It is well-known, at least in the compact case, that r-normal hyperbolicity is a persistent property under C^1 small perturbations of class C^r, that is, the persisting manifold is again r-normally hyperbolic and specifically C^r, see [HPS77, Thm. 4.1] or [Fen72, Thm. 2]. In other words, r-normal hyperbolicity is an 'open property' in the space of C^r systems with C^1 topology. Therefore, it is natural to only assume that the original manifold is C^r, but not smoother. Even if we start out with an r-NHIM $M \in C^\infty$, then after a perturbation we will generally only have a manifold $M_\varepsilon \in C^r$. We could, however, also have tried to obtain this manifold M_ε by first perturbing M to an intermediate manifold $M_{\varepsilon/2}$ and then perturb that manifold to M_ε. When applying a persistence theorem in the second step, we can only assume the initial manifold to be C^r.

This restricted C^r smoothness assumption forces us to be careful about the precise smoothness of each and every object. For example, a vector field on a C^r manifold can only be C^{r-1}. This could probably be overcome by considering discrete-time mappings instead of flows, but we need other smoothness improvements as well. For example, we want to model the persisting manifold as a section of the normal bundle

$N \in C^{r-1}$ of the original manifold, which is not smooth enough. So here we need a smoothing argument as well, cf. [Fen72, p 205].

We solve these problems by constructing an approximate, smoothed manifold $M_\sigma \in C^\infty$. This allows M to be modeled as a small section of the normal bundle N_σ of M_σ, so the system in a neighborhood of M can be transferred to N_σ while preserving smoothness and normal hyperbolicity properties. With this construction we need not worry about smoothness in the proof, while the conclusions are preserved up to C^r smoothness. Uniform estimates must be preserved though, so standard methods for constructing M_σ and N_σ do not readily apply or need a careful analysis.

First, we construct $\sigma > 0$ close approximations $M_\sigma \in C^\infty$ to M by globalizing a local chart construction of smoothing by convolution with a mollifier. Then we use the fact that M_σ has uniformly bounded 'second-order derivatives' to show that for a sufficiently small $\sigma > 0$, M_σ has a normal bundle diffeomorphic to a neighborhood $B(M; \delta) \subset Q$ of uniform size.

We recall some standard techniques on $\mathbb{R}^n$, see for example [Hör03, p 25]. Let $\varphi_\nu \in C_0^\infty(\mathbb{R}^n; \mathbb{R}_{\geq 0})$ be a mollifier function, with support in $B(0; \nu)$ and integral normalized to one for any $\nu > 0$. We also define a generic cut-off function $\chi_{\alpha,\beta} \in C^\infty(\mathbb{R}; [0, 1])$ such that

$$\chi_{\alpha,\beta}(x) = \begin{cases} 1 & \text{if } x \leq \alpha, \\ 0 & \text{if } x \geq \beta. \end{cases} \tag{2.31}$$

Note that $\varphi_\nu,\ \chi_{\alpha,\beta} \in C^k_{b,u}$ for any $k \geq 0$, as they are constant outside compact sets.

Lemma 2.34 (Smoothing by convolution) *Let $r,\ \delta r > 0$, $f \in C^{k\geq 0}_{b,u}\big(B(0; r + 2\,\delta r) \subset \mathbb{R}^m; \mathbb{R}^n\big)$, and fix $l > k$ and $\varepsilon > 0$. If the mollifier support radius $\nu > 0$ is chosen sufficiently small, then f can be approximated by a function $\tilde{f}$ such that*

1. *$\tilde{f} = f$ outside $B(0; r + \delta r)$;*
2. *$\tilde{f} \in C^l_{b,u} \cap C^\infty$ on $B(0; r)$ and wherever $f \in C^l_{b,u} \cap C^\infty$;*
3. *$\|\tilde{f} - f\|_k \leq \varepsilon$;*
4. *$\|\tilde{f}\|_l \leq C(\nu, l)\, \|f\|_0$ on $B(0; r)$, for some $C(\nu, l) > 0$.*

Note that $C(\nu, l)$ may grow unboundedly as $\nu \to 0$ or $l \to \infty$.

Proof A function that is C^{l+1}_b is automatically $C^l_{b,u}$, that is, uniformly continuous up to one degree less, so we only need to prove $\tilde{f} \in C^l_b \cap C^\infty$ for l shifted by one.

We construct $\tilde{f}$ by a combination of convolution and cut-off. Let $\hat{\chi}(x) = \chi_{r,r+\delta r}(\|x\|)$ for $x \in \mathbb{R}^m$ and define

$$\tilde{f}(x) = (1 - \hat{\chi}(x))f(x) + \hat{\chi}(x) \int_{\mathbb{R}^m} f(x - y)\, \varphi_\nu(y) \mathrm{d}y. \tag{2.32}$$

When $\nu < \delta r/2$, this $\tilde{f}$ is smooth on $B(0; r)$ and equal to f outside $B(0; r + \delta r)$.

The convolution approximates f in C^k-norm, as for any $0 \leq j \leq k$ and $x \in B(0; r + \delta r)$

$$\begin{aligned}
&\|\mathrm{D}^j(\varphi_\nu * f)(x) - \mathrm{D}^j f(x)\| \\
&\le \int_{B(0;\nu)} \|\mathrm{D}^j f(x-y) - \mathrm{D}^j f(x)\| \varphi_\nu(y)\mathrm{d}y \le \varepsilon_{\mathrm{D}^j f}(\nu),
\end{aligned}$$

so by uniform continuity of f up to kth derivatives, ν can be chosen small enough such that $\|(\varphi_\nu * f) - f\|_k \le \varepsilon$ on $B(0; r + \delta r)$. The map $x \mapsto \hat{\chi}(x)$ is $C^k_{b,u}$, so we can estimate for $j \le k$

$$\|\mathrm{D}^j \tilde{f}(x) - \mathrm{D}^j f(x)\| \le \sum_{i=0}^{j} \binom{j}{i} \|\mathrm{D}^i \hat{\chi}(x)\| \cdot \|\mathrm{D}^{j-i}(\varphi_\nu * f - f)(x)\| \le C_j\, \varepsilon(\nu).$$

Hence, we can construct $\tilde{f}$ close enough to f in C^k-norm by choosing ν small enough.

Uniform continuity of $\tilde{f}$ follows from uniform continuity of $\varphi_\nu * f$ as $\hat{\chi} \in C^k_{b,u}$ on its compact support. We find for $0 \le j \le k$

$$\begin{aligned}
&\|\mathrm{D}^j(\varphi_\nu * f)(x_2) - \mathrm{D}^j(\varphi_\nu * f)(x_1)\| \\
&\le \int_{B(0;\nu)} \|\mathrm{D}^j f(x_2 - y) - \mathrm{D}^j f(x_1 - y)\|\, \varphi(y)\mathrm{d}y \le \varepsilon_{\mathrm{D}^j f}(\|x_2 - x_1\|).
\end{aligned}$$

To estimate bounds for higher derivatives of $\tilde{f}$ within $B(0; r)$, we note that $\hat{\chi} = 1$ and let the derivatives act on φ_ν in the convolution: these are bounded on the compact domain of support, but bounds will depend on the size ν and degree l, while $\|f\|_0$ can be factored out. □

The smoothing technique in Lemma 2.34 is formulated for Euclidean space. To adapt it to manifolds in a uniform setting, we need to have uniformly sized coordinate charts, as well as uniform behavior of the function under these smoothing operations. We cannot simply use local coordinates and a partition of unity, because the images on different charts cannot be glued together on the target manifold. Instead, we will apply this smoothing operation sequentially on each coordinate chart in a cover. We require a cover that is locally finite with a global upper bound K on the number of charts covering a point, so that each point undergoes only a bounded number of smoothing operations and hence the final smoothed manifold M_σ differs by a controllable amount from the original M.

When the graph representation of M in one chart is modified, we need control on how much the graph is modified in overlapping charts. To this end, we extend Lemma 2.29 and Corollary 2.31.

Lemma 2.35 (Graph difference under coordinate transformations) *Let (Q, g) be a smooth Riemannian manifold of bounded geometry. Let $x_1, x_2 \in Q$ and let $\mathrm{T}_{x_i} Q = H_i \oplus V_i$, $i = 1, 2$ be splittings along horizontal and vertical perpendicular subspaces with* $\dim(H_1) = \dim(H_2)$. *Assume that $d(x_1, x_2) < \delta$ and that, for $i \neq j$,*

H_i is represented in normal coordinates at x_j by the graph of $L_i \in \mathcal{L}(H_j; V_j)$ with $\|L_i\| \le \zeta$.

Let $f_1, g_1 \in C^{k\ge 1}_{b,u}\big(B(0;\delta) \subset H_1; V_1\big)$ with $\|f_1\|_1, \|g_1\|_1 \le \varepsilon$ and $\|f_1\|_k, \|g_1\|_k \le C$.

When $\delta, \zeta, \varepsilon > 0$ are sufficiently small, then there exists a constant $\tilde{C}$ such that the graphs of f_1, g_1 are (partially) represented by functions $f_2, g_2 \in C^k_{b,u}(H_2; V_2)$ and $\|f_2 - g_2\|_k \le \tilde{C}\,\|f_1 - g_1\|_k$. This result is uniform for all $x_1, x_2 \in Q$.

Remark 2.36 The functions f_i, g_i may only be defined on parts of $B(x_i;\delta)$; all claims should thus be read as only for those points where the respective functions are defined. ◇

Proof Let $(\xi_i, \eta_i) \in H_i \oplus V_i$, $i = 1, 2$ denote normal coordinates, decomposed in the split directions at $x_i \in Q$. By Lemma 2.29, transformations between these coordinates are of the form (2.25), where $\|\tilde{\varphi}_{2,1}\|_{k+1}$ can be made uniformly small as $\delta, \zeta \to 0$.

We aim to apply the implicit function theorem to find a function f_2 on $B(x_2;\delta)$ whose graph corresponds to that of a function f_1 on $B(x_1;\delta)$. We define

$$\begin{aligned} &X = H_1 \times V_2, \quad Y = H_2 \times C^k_{b,u}(H_1; V_1), \quad Z = H_2 \times V_2, \quad \text{and} \\ &F : X \times Y \to Z : (\xi_1, \eta_2), (\xi_2, f_1) \mapsto \varphi_{2,1}(\xi_1, f_1(\xi_1)) - (\xi_2, \eta_2). \end{aligned} \tag{2.33}$$

Note that X and Z are isomorphic vector spaces, so we can apply the implicit function theorem with Y as parameter space. Moreover, if we have two functions f_1, f_2 whose graphs represent the same manifold on the intersection $B(x_1;\delta) \cap B(x_2;\delta)$, then we have

$$F\big(p_1 \circ \varphi_{2,1}^{-1}(\xi_2, f_2(\xi_2)), f_2(\xi_2), \xi_2, f_1\big) = 0$$

for all $\xi_2 \in H_2$ where this is defined, so the implicit function

$$G(\xi_2, f_1) = \big(p_1 \circ \varphi_{2,1}^{-1}(\xi_2, f_2(\xi_2)), f_2(\xi_2)\big)$$

encodes the representation f_2. We verify the conditions of the implicit function theorem:

$$\mathrm{D}_1 F\big((\xi_1, \eta_2), (\xi_2, f_1)\big) = \begin{pmatrix} O_H + p_1 \cdot \mathrm{D}\tilde{\varphi}_{2,1} \cdot (\mathbb{1} + \mathrm{D} f_1) & 0 \\ O_V \cdot \mathrm{D} f_1 & \mathbb{1} \end{pmatrix}$$

is unitary when $\tilde{\varphi}_{2,1} = f_1 = 0$. When δ, ε are sufficiently small, then these functions are still small enough such that $\mathrm{D}_1 F$ is invertible with uniformly bounded inverse, using Lemma A.1. Furthermore, $F \in C^k_{b,u}$, as the dependence on ξ_1, η_2, ξ_2 is clearly $C^k_{b,u}$, while the omega Lemma [AMR88, p 101] guarantees joint $C^k_{b,u}$-dependence on f_1 as well. Note that compactness of the domain of f_1 is not required, as we assume these functions to be uniformly bounded and thus have compact image.

The implicit function theorem has a corresponding formulation as a uniform contraction principle. The latter formulation shows that the implicit function G must be unique, while existence holds if $\tilde{\varphi}_{2,1}$, f_1, ξ_2 are sufficiently close to zero, due to a priori estimates. We apply Corollary A.4 as an extension of the implicit function theorem to conclude that $G \in C^k_{b,u}$. This means that $f_2 = p_2 \circ G(\,\cdot\,, f_1) \in C^k_{b,u}$ on suitable neighborhoods. Using formula (A.2) for DG, we can moreover conclude that f_2 depends Lipschitz on f_1. This follows from explicit control on the boundedness and continuity estimates, while variation with respect to f_1 only introduces additional $k+1$-order derivatives of $\tilde{\varphi}_{2,1}$, which can be assumed uniformly bounded. Hence, there exists some constant $\tilde{C}$ such that $\|g_2 - f_2\|_k \leq \tilde{C}\,\|g_1 - f_1\|_k$ and all estimates are uniform. □

Corollary 2.37 (Graph size under coordinate transformations) *Under the assumptions of Lemma 2.35, there exist constants A, B such that we have the estimate*

$$\|f_2\|_k \leq A\,\|f_1\|_k + B \tag{2.34}$$

on amplification of the size of a graph under coordinate transformations.

Proof We choose $g_1 = 0$ in Lemma 2.35 and set $A = \tilde{C}$. There exists a uniform bound B such that $\|g_2\|_k \leq B$, and when ζ, ε are sufficiently small, then for $\delta' = \frac{9}{10}\,\delta$ we have $B(0; \delta') \subset \mathrm{Dom}(g_2)$. Hence, we easily deduce

$$\|f_2\|_k \leq \|f_2 - g_2\|_k + \|g_2\|_k \leq \tilde{C}\,\|f_1\|_k + B$$

for all $x \in B(0; \delta')$ where f_1, f_2 are defined. □

Theorem 2.38 (Uniform smooth approximation of a submanifold) *Let $M \in C^{k\geq 1}_{b,u}$ be a uniformly immersed submanifold of a smooth Riemannian manifold (Q, g) of bounded geometry.*

Then for each $\sigma > 0$ and integer $l \geq k$, there exists a uniformly immersed submanifold $M_\sigma \in C^l_{b,u} \cap C^\infty$ and $\delta > 0$ such that $\|M_\sigma - M\|_k \leq \sigma$ with respect to normal coordinate charts of radius δ along both M and M_σ. If M is a uniformly embedded submanifold, then so is M_σ.

The proof relies on finding a (uniformly locally finite) cover of M and then in each chart make smooth the graph representation $h_\bullet$. All estimates are uniform, independent of the point $x \in M$, hence so is the final result. Smoothing is done sequentially in each chart, so we must be careful to check how smoothing in one chart influences the graph representation in other charts. This makes the technical estimates quite involved, but the basic idea is that we have uniform control on the size of changes in each $h_\bullet$ by the convolution kernel parameter ν in Lemma 2.34, as well as the size of this change in other charts.

Proof This proof contains a lot of interdependent size estimation parameters. Giving explicit choices and dependencies would clutter the proof needlessly, so we make a few remarks on beforehand. Any δ's denote sizes of normal coordinate balls and ε's are used for sizes of (changes in) functions in these coordinates. The parameter ν from Lemma 2.34 depends on most of the foregoing, while only the C^l bound (but not the C^k bound) of M_σ depends on the choice of ν. The various ε's will be fixed later, and depend on σ and global properties of M and (Q, g), but not on δ's. Also note that everything is independent of points $x \in M, Q$.

We fix $2\,\delta_1 = \delta_2 = \frac{1}{2}\delta_3$, $\varepsilon_\infty = 2\,\varepsilon_0$, and $C_\infty = C_0 + 1$ and choose $\delta_3, \zeta, \varepsilon_\infty, \varepsilon_\varphi > 0$ sufficiently small such that all the following statements hold true.

1. By Proposition 2.5 and Lemma 2.25, distances and metrics are $C = 2$ equivalent on balls $B(x; \delta_3)$, and d_Q, d_M are locally equivalent, all up to order $l + 1$.
2. By assumption of $M \in C^k_{b,u}$, we have for each $x \in M$ the representation $M_{x,\delta_3} = \operatorname{graph}(h_x)$ with $\|h_\bullet\|_1 \leq \varepsilon_0$ and $\|h_\bullet\|_k \leq C_0$.
3. By Corollary 2.26, there exists a uniformly locally finite cover $\bigcup_{i \geq 1} M_{x_i,\delta_2}$ of M with all x_i separated by at least δ_1, the balls M_{x_i,δ_1} already covering M, and bound K on the maximum number of M_{x_i,δ_2} intersecting any M_{x,δ_2}. Formula (2.10) shows that K depends on the ratio δ_2/δ_1, but does not increase when $\delta_3 \to 0$.
4. By Lemma 2.29, all coordinate transition maps $\varphi_{2,1}$ between any $x_1, x_2 \in Q$, $d(x_1, x_2) < \delta_3$ are C^l bounded. When the graph representations $H_i = \operatorname{Graph}(L_i)$ are bounded by $\zeta > 0$, then these are of the form $\varphi_{2,1} = O_H \oplus O_V + \tilde{\varphi}_{2,1}$, with $\|\tilde{\varphi}_{2,1}\|_{l+1} \leq \varepsilon_\varphi$. And by Corollary 2.31, this holds for the coordinate transformations between the x_i chosen for the cover of M.
5. By Lemma 2.35, there exists a constant $\tilde{C}$ that estimates the graph change under coordinate transformations from the previous point, when $\|f\|_1, \|g\|_1 \leq \varepsilon_\infty$ and $\|f\|_k, \|g\|_k \leq C_\infty$, while Corollary 2.37 holds on balls of size $\delta' = \frac{9}{10}\delta_3 > \delta_2$.
6. If M was a uniformly embedded submanifold, then let M_{x,δ_3} be the unique connected component of M in $B(x; \delta_3)$ for each $x \in M$.

Let M^j denote a modification of M after applying smoothing operations in the first j charts, and let h_i^j denote the graph representation of M^j in chart i. So, initially we have $M^0 = M$ and $h_i^0 = h_i$. Note that the sequence $\{h_i^j\}_{j \geq 1}$ is constant after some finite index $j(i)$, since it is changed at most K times by smoothing in overlapping charts. Thus, the final graphs are given by $h_i^\infty = h_i^{j(i)}$.

Initially, we have $\|h_\bullet\|_1 \leq \varepsilon_0$ and $\|h_\bullet\|_k \leq C_0$, and we assume that throughout the sequential smoothings it holds for all i, j that $\left\|h_i^j\right\|_1 \leq \varepsilon_\infty$ and $\left\|h_i^j\right\|_k \leq C_\infty$, and therefore the final h_i^∞ satisfy these estimates as well. Let ε_0 be sufficiently small, such that by a mean value theorem estimate

$$\|h_\bullet(\xi)\| = \|h_\bullet(\xi) - h_\bullet(0)\| \leq \|\mathrm{D}h_\bullet\|\,\|\xi\| \leq \varepsilon_0\,\|\xi\|,$$

we have $B(0; \delta_2) \subset \operatorname{Dom}(h_\bullet)$.

We apply the convolution smoothing Lemma 2.34 with some choice $r > \delta_1$ and $r + 2\,\delta r < \delta_2$ to sequentially make smooth h_i^{i-1} in the coordinate chart $B(x_i; \delta_3)$ to obtain $h_i^i \in C^l_{b,u} \cap C^\infty$ on $B(0; r)$. The $h_\bullet^j$ representations of the $M^j_{x_\bullet,\delta_3}$ overlap, so we must be careful that (at most) K repeated smoothing operations keep the $h_\bullet^j$ within the bounds required to apply this lemma, while at the same time we must ensure that each point on the sequence of manifolds M^j is smoothed to $C_{b,u}^l$ at some stage, even though M^j changes to $M_\sigma = M^\infty$ throughout the sequential smoothings.

Let us first show that each point is smoothed. We can keep track of each original point $x \in M$ as a sequence of points $x^j \in M^j$ throughout the smoothings, and once M^j is smoothed around x^j, then the convolution lemma guarantees that smoothness is preserved around the sequence x^j under further smoothing in other charts. Let $\Phi : M \xrightarrow{\sim} M_\sigma$ denote the diffeomorphism that assigns to $x \in M$ the final point $x^\infty \in M_\sigma$. Each point $x \in M$ is element of a graph h_i in at least one ball $B(x_i; \delta_1)$, so if the corresponding sequence of points $x^j \in M^j$ moves less than $r - \delta_1$, then it is smoothed in $B(x_i; r)$. Therefore, we choose ν in Lemma 2.34 small enough, such that

$$\|\tilde{f} - f\|_k \le \varepsilon(\nu) < \frac{r - \delta_1}{2\,K}.$$

The factor $2\,K$ accounts for at most K charts in which x^j is moved and $C = 2$ to correct for equivalence of distance in charts. Hence, the manifold is smoothed to $C_{b,u}^l \cap C^\infty$ at each point.

Next, we show that each h_i^j is defined at least on $B(0; r + 2\,\delta r)$ and satisfies the bounds ε_∞ and C_∞. Initially, we have $\|h_i\|_1 \le \varepsilon_0$, $\|h_i\|_k \le C_0$, and $\mathrm{Graph}(h_i) = M_{x_i,\delta_3}$ is well-defined in $B(x_i; \delta_3)$. So for $\varepsilon_0 \le \frac{1}{10}$, say, we must have $\|h_i\|_0 \le \frac{1}{10}\,\delta_3$ and $\mathrm{Dom}(h_i) \supset B(0; \frac{11}{10}\delta_2)$. The only reason that the domain of some h_i^j decreases is if either the graph moves outside of $B(0; \delta_3)$ or the modified manifold cannot be represented by a graph anymore. The latter cannot occur if $\mathrm{D}h_i^j$ stays bounded, while the former can be controlled by bounding $\|h_i^j\|_0$. Both can be controlled by estimating the C^1 changes $h_i^j - h_i^{j-1}$. First, in coordinate chart $i = j$ we can directly use the convolution smoothing Lemma 2.34 to conclude that $\|h_i^j - h_i^{j-1}\|_1 \le \varepsilon(\nu)$. In any other chart $i \ne j$, this change is amplified by a bounded factor $\tilde{C}$, as per Lemma 2.35, so we have

$$\|h_i^j - h_i^{j-1}\|_1 \le \tilde{C}\,\varepsilon(\nu).$$

When we choose ν small enough that

$$K\,\tilde{C}\,\varepsilon(\nu) < \min\Bigl(\varepsilon_0\,,\, \frac{1}{10}\,\delta_3\,,\, 1\Bigr)$$

holds, then this leads to

$$\begin{aligned}
\|h_i^j\|_0 &\le \|h_i\|_0 + K\,\tilde{C}\,\varepsilon(\nu) \le \frac{2}{10}\,\delta_3,\\
\|h_i^j\|_1 &\le \varepsilon_0 + \varepsilon_0 = \varepsilon_\infty,\\
\|h_i^j\|_k &\le \|h_i\|_k + K\,\tilde{C}\,\varepsilon(\nu) \le C_0 + 1 = C_\infty.
\end{aligned}$$

This shows that indeed the assumed bounds ε_∞ and C_∞ hold, and that h_i^j is defined at least on the ball $B(0; r + 2\,\delta r)$.

The sequential smoothings create and preserve $C_{b,u}^l \cap C^\infty$ smoothness, while every 'point' $x^j \in M^j$ is touched by these operations. Moreover, Lemma 2.34 and Corollary 2.37 together guarantee that the smoothing in each chart keeps

$$\|h_i^j\|_l \leq A\,\|h_j^{j-1}\|_0 + B \leq A\,C(\nu,l)\,\|h_j^{j-1}\|_0 + B \leq A\,C(\nu,l)\,\varepsilon_\infty + B$$

bounded with a uniform estimate, at least on charts $B(x_i; \delta_2)$.

Finally, we want to estimate the sizes and distance between the graphs $h_\bullet$, $h_\bullet^\infty$ in split coordinate charts of radius $\delta = \delta_2 - r$ along either M or M_σ. If x is a point either on M or M_σ, then it is contained in at least one ball $B(x_i; r)$ and $B(x; \delta) \subset B(x_i; \delta_2)$. If we also set $\varepsilon_\infty \leq \zeta$ and consider the coordinate transformation φ from normal coordinates at x_i to x, then Lemma 2.35 and Corollary 2.37 hold and can be used to estimate

$$\left\|h_x^\infty\right\|_l \leq A\left\|h_i^\infty\right\|_l + B \qquad \text{and} \qquad \left\|h_x^\infty - h_x\right\|_k \leq \tilde{C}\left\|h_i^\infty - h_i\right\|_k \leq \tilde{C}^2\,K\,\varepsilon(\nu)$$

for all points in the domains of h_x and $h_x^\infty - h_x$ within $B(0; \delta)$. So if we set $\tilde{C}^2\,K\,\varepsilon(\nu) < \sigma$, then M_σ is C^k close to M in normal coordinate charts of radius δ along either M or M_σ, while at the same time $M_\sigma \in C_{b,u}^l \cap C^\infty$.

If M is a uniformly embedded submanifold, then δ_3 was chosen small enough such that M_{x,δ_3} is the unique connected component of M in $B(x; \delta_3)$ for any $x \in M$. We now show that the same holds for M_σ with balls of radius δ. Let $\tilde{x} \in M_\sigma$ be arbitrary and $x = \Phi^{-1}(\tilde{x}) \in M$. We take $\tilde{y} \in M_\sigma \cap B(\tilde{x}; \delta)$ and want to prove that $\tilde{y} \in (M_\sigma)_{\tilde{x},\delta}$. By the uniform estimates made before, both M_{x,δ_2} and $(M_\sigma)_{\tilde{x},\delta_2}$ can be represented by graphs h_x, h_x^∞ respectively in coordinates $B(0; \delta_2) \subset \mathrm{T}_x M \oplus N_x$. We have $y = \Phi^{-1}(\tilde{y}) \in B\big(\tilde{x}; \delta + (r - \delta_1)\big) \subset B\big(x; \delta + 2(r - \delta_1)\big)$, so $y \in B(x; \delta_2) \cap M = M_{x,\delta_2} = \mathrm{Graph}(h_x)$.

By the construction of M_σ we have $x \in B(x_i; r)$ in some chart i, but also $(M_\sigma)_{x,\delta_2} = \mathrm{Graph}(h_x)$. Let $y \in B(x; \delta) \cap M_\sigma$; we want to prove that $y \in (M_\sigma)_{x,\delta} = \mathrm{Graph}(h_x^\infty)$. Since $y \in B(x; \delta)$, then for its original it must hold that

$$y^0 \in B\big(x; \delta + (r - \delta_1)\big) \subset B\big(x_i; r + \delta + (r - \delta_1)\big),$$

hence $y^0 \in M_{x_i, r+\delta+(r-\delta_1)} = \mathrm{Graph}(h_x)$. Following the change of M to M_σ in coordinates around x, we see that $y \in \mathrm{Graph}(h_x^\infty)$ must hold. □

2.5 Embedding into a Trivial Bundle

Let $\pi : N \to M$ be the normal bundle over M immersed in (Q, g), a Riemannian manifold of bounded geometry. We are going to construct a trivial bundle $\overline{N}$ over M that contains N and preserves uniform properties. As a second step, we extend a normally hyperbolic vector field to this trivial bundle setting. This procedure is also alluded to in [Sak94, p. 333–334], but especially in the case of bounded geometry requires a more careful inspection.

Theorem 2.39 (Uniform embedding of a normal bundle in a trivial bundle) *Let $M \in C^{k \geq 1}_{b,u}$ be a uniformly immersed submanifold of the bounded geometry manifold (Q, g). Then there exists an embedding $\lambda : N \hookrightarrow \bar{N}$ of the (nontrivial) normal bundle $\pi : N \to M$ into a larger, trivial vector bundle $\bar{N} = M \times \mathbb{R}^{\bar{n}}$. The embedding map $\lambda \in C^{k-1}_{b,u}$ is a quasi-isometry when restricted to $N_{\leq \eta}$ for any $\eta > 0$ and the splitting $\bar{N} = \lambda(N) \oplus N^{\perp}$ is $C^{k}_{b,u}$, where $N^{\perp}$ is chosen perpendicular to $\lambda(N)$ according to the standard Euclidean metric on $\mathbb{R}^{\bar{n}}$.*

Note that λ only has smoothness C^{k-1} since $N \in C^{k-1}$ is the normal bundle of $M \in C^{k}$. The image bundle $\lambda(N) \subset \bar{N}$ has smoothness C^{k}, though, since its construction only involves the immersion $M \to Q$. This increase of smoothness is possible because we do not view $\lambda(N)$ as a normal bundle with respect to the differentiable structure of Q anymore. This can be compared to the remark in [Fen72, p. 205] and the reference to [Whi36, Lem. 23] therein.

The idea of the proof is to use normal coordinate charts of Q covering M to construct local trivialization maps of N. In such charts $B(x; \delta)$, we have, from Definition 2.21, $M_{x,\delta} = \exp_x\big(\mathrm{Graph}(h_x)\big) \subset M$ and trivialization maps τ_x for the vector bundle trivialization diagram

$$\begin{array}{ccc} N \supset N|_{M_{x,\delta}} & \xrightarrow{\ \tau_x\ } & M_{x,\delta} \times N_x \\ \Big\downarrow \pi & \swarrow p_1 & \\ M \supset M_{x,\delta} & & \end{array} \tag{2.35}$$

Then we take a uniformly locally finite cover of M by sets $M_{x_i,\delta}$. The trivializations on each $M_{x_i,\delta}$ induce a spanning set of sections, i.e. a frame. Using the uniformity of the cover, we can globally glue these frames together to obtain $\mathrm{rank}(\bar{N}) = \bar{n} = (K+1)\,\mathrm{rank}(N)$. Here, K is the maximum number of overlapping charts in the cover. This identifies $\lambda(N)$ as the subbundle of $\bar{N} = M \times \mathbb{R}^{\bar{n}}$ spanned by these glued frames.

Proof Let δ be Q-small as in Definition 2.8, as well as sufficiently small such that M is given as the graph of $h_{\bullet} : \mathrm{T}_{\bullet}M \to N_{\bullet}$ in normal coordinates as in Definition 2.21. For any $x \in M$ we have a trivialization map

$$\tau_x = (\exp_x, p_2) \circ \mathrm{D}\exp_x^{-1} : N|_{M_{x,\delta}} \xrightarrow{\sim} M_{x,\delta} \times N_x, \tag{2.36}$$

where we canonically identified $\mathrm{T}(\mathrm{T}_x Q) \cong (\mathrm{T}_x Q)^2$ and apply $\exp_x$ only on the base $\mathrm{T}_x Q$ and p_2 on the fibers of $\mathrm{T}(\mathrm{T}_x Q)$. In a normal coordinate representation (see page 52, Fig. 2.2 on the right) this just means that we project the normal fiber $N_{x'}$ at any point $x' = \exp_x(\xi, h_x(\xi)) \in M_{x,\delta}$ onto N_x. By Corollary 2.31 this projection is approximately orthogonal and bounded away from non-invertibility for small δ, hence $\tau_\bullet \in C^{k-1}_{b,u}$ and it is a quasi-isometry, but only on a finitely sized neighborhood $N_{\leq\eta}|_{M_{x,\delta}}$ since it acts linearly on the fibers of $N|_{M_{x,\delta}}$. We then choose a uniformly locally finite cover $\bigcup_{i\geq 1} M_{x_i,\delta}$ of M such that the sets $M_{x_i,\delta/2}$ already cover M.

Next, we prove the existence of a finite set of $C^{k-1}_{b,u}$ sections that everywhere span N. Let $G = (V, E)$ be the (possibly infinite) graph whose vertices are sets in the cover, $V = \{M_{x_i,\delta}\}_{i\geq 1}$, and edges are added between overlapping sets, i.e. $E = \{(A, B) \in V^2 \mid A \cap B \neq \emptyset\}$. Each set in the cover overlaps at most K other sets, so the maximal degree of G is bounded by K. Therefore, we can 'color' the vertices of G with numbers $\{0, \ldots, K\}$ such that no two connected vertices have the same number. Sequentially for each $i \geq 1$, set the number of vertex i to one of the numbers $\{0, \ldots, K\}$ that is not already taken by its neighbors. We thus obtain a map $c : \mathbb{N} \to \{0, \ldots, K\}$ such that each preimage $c^{-1}(k)$ labels a collection of mutually disjoint sets of the cover.

Let n denote the rank of N and let $\bar{N} = M \times \mathbb{R}^{\bar{n}}$ be a trivial bundle with rank $\bar{n} = n\,(K+1)$. On each $M_{x_i,\delta}$ we have an orthogonal frame e_i of n sections that span N, induced by the local trivialization, while on $\bar{N}$ we have the global orthogonal frame $\bar{e}$ of standard unit sections. The latter can also be viewed as a $(K{+}1)$-tuple of n-frames $\{\bar{e}_k\}_{0\leq k\leq K}$ on $\mathbb{R}^{\bar{n}} = (\mathbb{R}^n)^{K+1}$. Since all spaces have (the standard Euclidean) inner products, the dual frames can be canonically identified as the inverse $e_i^* = e_i^{-1} : N_{x_i} \to \mathbb{R}^N$ and a projection $\bar{e}_k^* = p_k : \mathbb{R}^{\bar{n}} \to \mathbb{R}^n$ onto the k-th n-tuple of all $\bar{n}$ coordinates respectively. Let the functions χ_i be a square-sum partition of unity subordinate to the cover according to Corollary 2.18 and define the embedding

$$\lambda : N_{\leq\eta} \hookrightarrow \bar{N}_{\leq\eta} : (m, \nu) \mapsto \sum_{i\geq 1} \chi_i(m)\, \bar{e}_{c(i)}\, e_i^*\, \tau_{x_i}(m, \nu). \tag{2.37}$$

This mapping is $C^{k-1}_{b,u}$ as a composition of such maps (and can be extended, albeit non-boundedly so, to a map $N \hookrightarrow \bar{N}$). The τ_{x_i} are quasi-isometries, while $\bar{e}_{c(i)}\, e_i^*$ is isometric on each $M_{x_i,\delta}$. Each frame $\bar{e}_{c(i)}$ is orthogonal to the frame of any overlapping set $M_{x_j,\delta}$ since $c(j) \neq c(i)$ and the χ_i squared sum to one. Thus $\sum_i \chi_i\, \bar{e}_{c(i)}\, e_i^*$ is an isometry, and so λ is a quasi-isometry.

Let p and $p^\perp = \mathbb{1} - p$ denote the projections from $\bar{N}$ onto N and $N^\perp$, respectively. One can verify that

$$p = \sum_{i,j\geq 1} \chi_i\, \chi_j\, \bar{e}_{c(i)}\, \bar{e}^*_{c(j)} \tag{2.38}$$

is the projection onto N by noting that $\lambda(N)$ equals the image of $\sum_{i\geq 1} \chi_i\, \bar{e}_{c(i)}$, while the identities

$$\bar{e}_k^*\, \bar{e}_l = \delta_{kl} \quad \text{and} \quad \sum_{i\geq 1} \chi_i^2 = 1$$

can be used to show that $p^2 = p$. Formula (2.38) shows that both $p, p^\perp \in C^k_{b,u}$, hence the splitting $\bar{N} = \lambda(N) \oplus N^\perp$ is $C^k_{b,u}$. □

Next, we must extend a vector field v on N to the larger bundle $\bar{N}$. There are additional directions along the fibers of $N^\perp$ and the extended vector field $\bar{v}$ must be such that it is normally hyperbolic in these directions as well. On the other hand, the uniform boundedness of v must be preserved. We do not assume here that M is the exact invariant manifold, since these results shall be applied after application of Theorem 2.38, which has smoothed and slightly altered M such that it is not the original NHIM anymore. The extension $\bar{v}$ will keep N invariant and is identical to v on N, so in the end, we can conclude that the perturbed manifold is contained in N and restrict to the original setting again.

Lemma 2.40 **(Normally hyperbolic extension of a vector field)** *Let $\lambda : N \hookrightarrow \bar{N} = M \times \mathbb{R}^{\bar{n}}$ be a trivializing embedding of vector bundles as in Theorem 2.39 and let $v \in C^{l,\alpha}_{b,u}$ be a vector field on N with $l + \alpha \leq k - 2$. Let $\bar{N}|_{\leq\eta}$ be the restriction to some radius $\eta > 0$. Then v can be extended to a vector field $\bar{v}$ on $\bar{N}$, such that $\bar{v}$ is $C^{l,\alpha}_{b,u}$ on $\bar{N}_{\leq\eta}$, leaves N invariant, and contracts at a given exponential rate $\rho < 0$ along the fiber direction of $N^\perp$ towards $N \subset \bar{N}$.*

To extend v to a vector field $\bar{v}$ on $\bar{N}$ with the required properties, we must do two things. First of all, v must be extended from N through λ to the whole of $\lambda(N) \oplus N^\perp$ and secondly, a normal component along the fibers of $N^\perp$ must be added to make $\bar{v}$ contracting, thus normally hyperbolic in that direction. The idea can be expressed in local coordinates $(m, y, z) \in \lambda(N) \oplus N^\perp$ as

$$\bar{v}(m, y, z) = \hat{v}(m, y, z) + v^\perp(m, y, z) = \big(v(m, y), \rho\, z\big),$$

where $\hat{v}(m, y, z) = v(m, y)$ points 'horizontally' along $\lambda(N)$ and $v^\perp(m, y, z) = \rho\, z$ is the 'vertical' component along the $N^\perp$ fibers. By construction, the latter has the required contraction property in the $N^\perp$ direction, while it preserves $\lambda(N) \oplus \{0\}$ as an invariant manifold. We shall make this intuitive idea rigorous by introducing an appropriate bounded connection to lift v to $\hat{v}$ for $z \neq 0$; the second term $v^\perp$ is canonically defined.

Proof The embedding map λ is a quasi-isometry and of class $C^{k-1}_{b,u}$, hence the pushforward $\lambda_*(v) = \mathrm{D}\lambda \cdot v \circ \lambda^{-1}$ is a $C^{l,\alpha}_{b,u}$ vector field on $\lambda(N) \subset \bar{N}$. From now on we identify N with $\lambda(N)$ as well as v with its pushforward.

Let g be the standard Euclidean metric on $\bar{N}$ and ∇ the compatible, trivial, flat connection. The restricted metric $g^\perp = g|_{N^\perp}$ is preserved by the connection $\nabla^\perp = p^\perp \cdot \nabla$. We create the pullback bundle

$$E = \pi_N^*(N^\perp) = \Big\{(y, z) \in N \times N^\perp | \pi_N(y) = \pi_{N^\perp}(z)\Big\} \cong N \oplus N^\perp = \bar{N}.$$

Note that we identify this pullback of $N^\perp$ along the projection $\pi_N : N \to M$ with the vector bundle $p : \overline{N} \to N$, that is, we view N as the base manifold and $\overline{N}$ as bundle over N with fibers $\pi_N^{-1}(y) = N^\perp_{\pi_N(y)}$. We naturally endow E with the pullback connection $\hat{\nabla} = \pi_N^*(\nabla^\perp)$. With this connection, v can be lifted to a unique vector field $\hat{v} \in \mathfrak{X}(\overline{N})$ that is horizontal along N on $E \cong \overline{N}$, and thus the flow of $\hat{v}$ preserves the norm along the fibers of E, that is, $\hat{v} \cdot \| \cdot \|_E = 0$. More heuristically, we can say that the pullback $\pi_N : N \to M$ introduces a trivial additional base coordinate $y \in N_m$ to the bundle $\pi_{N^\perp} : N^\perp \to M$.

To prove that $\hat{v} \in C^{l,\alpha}_{b,u}$, we first recover an explicit representation of the Christoffel symbols of $\hat{\nabla}$ in terms of trivial coordinates on $E \cong \overline{N} = M \times \mathbb{R}^{\bar{n}}$, and then a representation for the lifted vector field $\hat{v}$. Let $s \in \Gamma(\overline{N})$, $s \equiv s_0 \in \mathbb{R}^{\bar{n}}$ be a constant section. Let $m \in B(m_0; \delta)$ denote normal coordinates in M. Define $s^\perp = p^\perp \cdot s \in \Gamma(N^\perp)$ and $\hat{s} = \pi_N^*(s^\perp) \in \Gamma(E)$. Let $\hat{X} \in \mathrm{T}N$ a tangent vector in the base of E and $X = \mathrm{D}\pi_N(\hat{X}) \in \mathrm{T}M$. Then we find for the covariant derivative $\hat{\nabla}$ on E

$$\hat{\nabla}_{\hat{X}}\, \hat{s} = \pi_N^*\big(p^\perp \cdot \nabla_X(p^\perp \cdot s)\big) = \pi_N^*\left(p^\perp \cdot X^i\Big[p^\perp \frac{\partial s}{\partial m^i} + \frac{\partial p^\perp}{\partial m^i}\, s\Big]\right).$$

We read off that the Christoffel symbols are given by $p^\perp \frac{\partial p^\perp}{\partial m^i}\,(\mathrm{D}\pi_N)^i$, so they are $C^{k-1}_{b,u}$. The horizontal lift

$$\hat{v}(m, y, z) = v(m, y) - p^\perp \cdot \Big(\frac{\partial p^\perp}{\partial m^i}\, z\Big)(\mathrm{D}\pi_N\, v(m, y))^i, \tag{2.39}$$

then, is also $C^{l,\alpha}_{b,u}$ since all functions involved are at least $C^{l,\alpha}_{b,u}$ in these coordinates, and z is bounded on $\overline{N}_{\leq\eta}$.

We define $v^\perp$ as the Euler vector field along the fibers of $\overline{N} \cong E$, taking values in $\mathrm{Vert}(E)$. Each fiber $E_{(m,y)} = N^\perp_m$ is a linear space, so the tangent space at any point is canonically identified with the fiber itself, which allows us to canonically define

$$v^\perp : \overline{N} \to \mathrm{T}\overline{N} : (m, y, z) \mapsto \rho\, z \in \mathrm{T}_z N^\perp_m = \mathrm{Vert}(E)_{(m,y,z)}. \tag{2.40}$$

This vector field leaves $N \subset \overline{N}$ invariant, while generating a flow that attracts towards N at the exponential rate $\rho < 0$. It is clear that $v^\perp \in C^k_{b,u}$ for any k when z is bounded on $\overline{N}_{\leq\eta}$.

We conclude that the vector field $\bar{v} = \hat{v} + v^\perp$ indeed leaves N invariant. Since $\hat{v}$ is neutral in the fiber directions of E and $v^\perp$ contracting at rate ρ, it follows that $\bar{v}$ is contracting with rate ρ as well. The combined vector field $\bar{v}$ is defined in terms of v and other functions that are all at least $C^{l,\alpha}_{b,u}$, hence $\bar{v} \in C^{l,\alpha}_{b,u}$. □

2.6 Reduction of a NHIM to a Trivial Bundle

Having set up the theory of bounded geometry spaces, we are finally in a position to reduce a general normally hyperbolic system to the setting of a trivial bundle. That setting is required to apply our basic Theorem 3.2 on persistence of NHIMs. Let $M \in C^{k,\alpha}_{b,u}$ with $k \geq 2$ and $0 \leq \alpha \leq 1$ be a uniformly immersed or embedded submanifold in (Q, g) of bounded geometry and furthermore assume that M is an r-NHIM with $r = k + \alpha$ for the vector field $v \in C^{k,\alpha}_{b,u}$ on Q.

Remark 2.41 The bounded smoothness requirement $k \geq 2$ is dictated by Theorem 2.33. It is not present in the compact case where the normal bundle can be "jiggled slightly" [Fen72, Prop. 2] to make it sufficiently smooth to model a C^k flow for $k \geq 1$. Hypotheses 2 and 3 in [BLZ99, p. 987] require similar conditions in the noncompact setting in Banach spaces, see also the discussion after Corollary 3.5 and Remark 3.13. I have not investigated in detail whether the C^2_b requirement is necessary, or if $M \in C^1_{b,u}$ could be modeled as a sufficiently small section of the normal bundle of a smooth approximate manifold. ◇

We reduce this system to a trivial bundle in the following steps:

1. approximate M by a smoothed manifold M_σ;
2. construct a tubular neighborhood of M in the normal bundle N of M_σ;
3. embed N into a trivial bundle $\bar{N} = M_\sigma \times \mathbb{R}^{\bar{n}}$, and construct an extended, normally hyperbolic vector field $\bar{v}$;
4. after application of the basic persistence theorem in the enlarged bundle, push the results to the original setting and conclude that M persists.

Proof of Theorem 3.1. Assume the hypotheses of the theorem. First, Theorem 2.38 gives a smooth, C^k-close approximation $M_\sigma \in C^l_{b,u}$ of M, where the choice $l = k + 10$ suffices and $\sigma > 0$ will be fixed later. The $C^{k,\alpha}$ bounds of M_σ are uniformly close to those of M for all σ small. Then, Theorem 2.33 says that there exists a tubular neighborhood $\varphi : N_{\leq \eta} \xrightarrow{\sim} B(M_\sigma; \eta)$ where the size $\eta > 0$ depends only on the C^2 bounds of M_σ. These bounds are of the same order as those of M, independent of σ. Hence, we can choose σ so small that $\|M_\sigma - M\|_k \leq \sigma \leq \frac{\eta}{2}$ and the neighborhood $B(M; \eta/2)$ is fully within the tubular neighborhood $N_{\leq \eta}$ of M_σ. The map φ is a C^{l-1} bounded (local) diffeomorphism and a quasi-isometry, so by the reasoning[4] before Definition 2.32, a pullback by φ does not change the normal hyperbolicity growth rates of the vector field v. The bounded continuous splitting (1.9) of $\mathrm{T}_M Q$ is also preserved.

Next, as a result of Theorem 2.39, N is embedded into the trivial bundle

$$\lambda : N \hookrightarrow \bar{N} = M_\sigma \times \mathbb{R}^{\bar{n}} = N \oplus N^\perp.$$

[4] This is similar to Fenichel's argument in [Fen72, p 203] that normal hyperbolicity is independent of a choice of metric when M is compact.

The embedding is a quasi-isometry and so preserves hyperbolicity properties of v. The extended vector field $\bar{v} \in C^{k,\alpha}_{b,u}$ on $\overline{N}_{\leq\eta}$ is constructed in Lemma 2.40 as a lift of v, so the flow preserves $N_{\leq\eta}$ and intertwines with the projection onto N, while in the perpendicular direction along the fibers of $N^{\perp}$ it has the same normal hyperbolicity properties as v. Boundedness of the invariant splitting $\mathrm{T}_M Q = \mathrm{T}M \oplus N^+ \oplus N^-$ is also preserved under these quasi-isometries. The additional directions along $N^{\perp}$ are stable and invariant, and have bounded projections by construction. Thus, M is an r-NHIM for $\bar{v}$ as well.

The invariant manifold M is given by the graph of a section $h \in \Gamma(N) \subset \Gamma(\overline{N})$, and from Theorem 2.38 it follows that $\|h\|_k \leq \sigma$ while $h \in C^{k,\alpha}_{b,u}$. By Lemma 2.27, $X = M_\sigma$ has bounded geometry of order $l-2 = k+8$, which is sufficiently smooth for the conditions of Theorem 3.2, while $Y = \mathbb{R}^{\bar{n}}$ is clearly a Banach space. Hence, we are in the trivial bundle setting, and all conditions are satisfied.

A small perturbation of v in the original setting in Q corresponds to a small perturbation of $\bar{v} \in \mathfrak{X}(\overline{N})$, while $N_{\leq\eta}$ is preserved under the flow by construction. Therefore, after application of Theorem 3.2 we recover a unique persistent invariant manifold $\tilde{M}$ and by construction $\tilde{M} = \mathrm{Graph}(\tilde{h}) \subset N$, so we can restrict the system to N. Then it can be transferred back to Q under the quasi-isometries of the embedding $\lambda : N \hookrightarrow \overline{N}$ and the tubular neighborhood map $\varphi : N_{\leq\eta} \to B(M;\eta) \subset Q$.

All size estimates can be transferred between the settings (with bounded factors) due to the near isometry and uniform $C^{k,\alpha}$ boundedness of φ and λ. We conclude that $\tilde{M}$ is a $C^{k,\alpha}_{b,u}$ submanifold of Q for appropriate estimates σ, ε, and δ, where σ must be chosen sufficiently small as well to make $\|\tilde{M} - M\|_{k-1}$ small.

This completes the reduction from the general setting in bounded geometry to that of a trivial bundle and proves Theorem 3.1. □

Chapter 3
Persistence of Noncompact NHIMs

This chapter contains the main proof of persistence of noncompact normally hyperbolic invariant manifolds, formulated in Theorem 3.2. This theorem is formulated in a specific setting: we assume that the invariant manifold M is (nearly) the zero section of a trivial vector bundle. This is a slightly more general formulation than in [Hen81; Sak90]. There, it is assumed that in a product $X \times Y$ of Euclidean (or Banach) spaces, the invariant manifold M is given as the graph of a function $h : X \to Y$. We shall also assume that Y is a vector space, but we let X instead be a Riemannian manifold that is finite-dimensional and has bounded geometry. In Chap. 2 on bounded geometry, we extended the result obtained here to a setting where M is assumed to be a general submanifold of a Riemannian manifold (Q, g) that is again finite-dimensional and of bounded geometry. We assume the basic statements from Sect. 2.1 to be known.

This chapter is organized as follows. First we state the two main theorems; both the general version with M a submanifold of Q and the $X \times Y$ trivial bundle version to be proved in this chapter. We provide detailed remarks on these theorems and compare them to the literature. Then we present an outline of the proof of Theorem 3.2. Section 3.3 presents some thoughts on replacing the classical compactness by uniformity conditions, and presents examples that indicate the necessity of various assumptions we impose.

In Sect. 3.4 we transform the (still somewhat geometrical) formulation of Theorem 3.2 into a more explicit setup suitable for analysis. In the subsequent section, we prove (with *relatively* little work) the existence and uniqueness of the persistent manifold $\tilde{M} = \mathrm{Graph}(\tilde{h})$. It automatically follows that $\tilde{h}$ is bounded and uniformly Lipschitz.

The last sections are devoted to the tougher job of proving $C^{k,\alpha}$ smoothness, exhausting the spectral gap. A formal scheme is set up, and we work out the details for C^1 smoothness. Higher, C^k smoothness follows along the same lines by induction. The addition of Hölder continuity to obtain $C^{k,\alpha}$ smoothness is included as a natural extension to (uniform) continuity that slightly simplifies the spectral gap estimates. See the proof outline and the introduction of Sect. 3.7 for more details.

J. Eldering, *Normally Hyperbolic Invariant Manifolds*, Atlantis Series in Dynamical Systems 2, DOI: 10.2991/978-94-6239-003-4_3,

3.1 Statement of the Main Theorems

The main theorem on persistence was already formulated in the introduction. We state it again to directly compare it to the trivialized bundle version of Theorem 3.2. The main theorem is reduced to this trivialized version in Sect. 2.6; in this chapter we shall prove the latter version. Then we formulate corollaries of these theorems, both to present simpler versions and to compare our result to well-known results from the literature.

Theorem 3.1 **(Persistence of noncompact NHIMs in bounded geometry)** *Let $k \geq 2, \alpha \in [0, 1]$ and $k + \alpha$. Let (Q,g) be a smooth Riemannian manifold of bounded geometry and $v \in C^{k,\alpha}_{b,u}$ a vector field om Q. Let $M \in C^{k,\alpha}_{b,u}$ be a connected, complete submanifold of Q that is r-normally hyperbolic for the flow defined by v, with empty unstable bundle, i.e. rank $(E^+) = 0$.*

Then for each sufficiently small $\eta > 0$ there exists a $\delta > 0$ such that for any vector field $\tilde{v} \in C^{k,\alpha}_{b,u}$ with $||\tilde{v} - v||_1 < \delta$ there is a unique submanifold $\tilde{M}$ in the η -neighborhood of M, such that $\tilde{M}$ is diffeomorphic to M and invariant under the flow defined by $\tilde{v}$ Moreover $\tilde{M}$ is $C^{k,\alpha}_{b,u}$ and the distance between $\tilde{M}$ and M can be made arbitrarily small in C^{k-1}-norm by choosing $||\tilde{v} - v||_{k-1}$ sufficiently small.

Theorem 3.2 **(Persistence of noncompact NHIMs in a trivial bundle)** *Let $k \geq 2$, $\alpha \in [0, 1]$ and $r = k + \alpha$. Let (X, g) be a smooth, complete, connected Riemannian manifold of bounded geometry and Y a Banach space. Let $v_\sigma \in C^{k,\alpha}_{b,u}$ be a family of vector fields defined on a uniformly sized neighborhood of the zero-section in $X \times Y$ with family parameter $\sigma \in (0, \sigma_0]$. Let the submanifold $M_\sigma = \mathrm{Graph}(h_\sigma)$ be given as the graph of a function $h_\sigma \in C^{k,\alpha}_{b,u}(X; Y)$ and let M_σ be an r-NHIM with* $\mathrm{rank}(E^+) = 0$ *for the flow defined by v_σ where all estimates are uniform in σ and additionally $\|h_\sigma\|_2 \leq \sigma$ holds.*

Then for each sufficiently small $\eta > 0$ there exist $\sigma_1, \delta > 0$ such that for any $\sigma \in (0, \sigma_1]$ and any vector field $\tilde{v} \in C^{k,\alpha}_{b,u}$ with $\|\tilde{v} - v_\sigma\|_1 < \delta$, there is a unique submanifold $\tilde{M} = \mathrm{Graph}(\tilde{h})$, $\tilde{h} : X \to Y$, $\left\|\tilde{h}\right\|_0 \leq \eta$ such that $\tilde{M}$ is invariant under the flow defined by $\tilde{v}$. Moreover, $\tilde{h} \in C^{k,\alpha}_{b,u}$ and $\left\|\tilde{h}\right\|_{k-1}$ can be made arbitrary small by choosing $\|h\|_k$ and $\|\tilde{v} - v_\sigma\|_{k-1}$ sufficiently small.

Remark 3.3 Let us make some remarks on these theorems.

1. The spectral gap condition contained in Definition 1.11 of r-normal hyperbolicity is essential to the proof. The $C^{k,\alpha}$-smoothness result is optimal, see Sect. 1.2.1.
2. In Theorem 3.1, both M and $\tilde{M}$ are assumed to be (non-injectively) immersed according to Definition 2.21. If M is assumed uniformly embedded according to Remark 2.22, then $\tilde{M}$ will be uniformly embedded again when δ is sufficiently small.
3. The additional family parameter σ in Theorem 3.2 is required to reduce Theorem 3.1 to this case. If the unperturbed manifold is given as $M = X \times \{0\}$, i.e. the zero

section, then the family v_σ can simply be taken constant and all σ dependence can be dropped from the formulation.

4. We only obtain a C^{k-1}-norm estimate for the perturbation distance of $\tilde{M}$ away from M, even though $\tilde{M} \in C^{k,\alpha}$ is preserved. This is due to a linearization along Y and the smoothing convolution used to restore $C^{k,\alpha}$ smoothness after linearization. See Sect. 3.4, in particular Remarks 3.13 and 3.15, for more details. I fully expect it to hold that $\tilde{M}$ and M are $C^{k,\alpha}$ close when $\|\tilde{v} - v\|_{k,\alpha}$ is small.
5. The minimum smoothness requirement $k \geq 2$ is a stronger assumption than $k \geq 1$ in the well-known compact case. This seems to be intrinsic to the noncompact case. If the spectral gap condition only holds for some $1 \leq r < 2$, then we can still obtain a perturbed manifold $\tilde{M}$. This manifold $\tilde{M}$ will generally not have better than C^r smoothness, though.
6. We allow both values $\alpha = 0$ and $\alpha = 1$, where $\alpha = 0$ is considered an empty condition (besides the boundedness and uniform continuity). Thus, if $r = k + 1$ satisfies the spectral gap condition (1.11), then we can choose both $C^{k,1}$ or $C^{k+1,0}$ as resulting smoothness for $\tilde{M}$, if M had the same smoothness. Thus, if M was sufficiently smooth, then the choice $\tilde{M} \in C^{k+1,0}$ yields the best result. Note, though, that by Rademacher's theorem, Lipschitz functions are differentiable almost everywhere, so the difference is not that big.
   Finally, it should also be noted that the spectral gap condition is a strict inequality on r, so if we can choose $r = k$ integer, then we can also find an $\alpha > 0$ such that $r' = k + \alpha$ satisfies the spectral gap as well. This shows that in this context $C^k_{b,u}$ 'integer' smoothness really is a special case of $C^{k,\alpha}_{b,u}$ 'fractional' smoothness.
7. Both Riemannian manifolds Q in Theorem 3.1 and X in Theorem 3.2 are assumed to be finite-dimensional; multiple results on bounded geometry crucially depend on this fact. On the other hand, we allow Y to be an infinite-dimensional Banach space simply because everything naturally generalizes to that setting. Note that we do not allow semi-flows as in [Hen81; BLZ08], so the case that Y is infinite-dimensional may not be that useful.
8. These results are weaker than those in the well-known compact case in a few aspects. First of all, we use a stricter notion of normal hyperbolicity, see Remark 1.10. This seems to be a fundamental restriction of the Perron method; the more general definition of normal hyperbolicity is successfully applied to noncompact manifolds in [BLZ08]. Secondly, we only include the stable normal bundle E^-. Adding the unstable bundle E^+ as well should be possible, see Sect. 4.4 for more details.
9. While we do prove that the NHIM persists into a new invariant manifold $\tilde{M}$, we do not prove that $\tilde{M}$ is again normally hyperbolic. I fully expect this to be true though: the perturbed flow satisfies slightly perturbed exponential growth conditions and the spectral gap is an open condition, so should be preserved under sufficiently small perturbations. It remains to be proven that $\tilde{M}$ again has a continuous invariant splitting (1.19) with bounded projections. This is one possible reason for breakdown of normal hyperbolicity [HL06]. In Sect. 4.5 we sketch how to recover the invariant stable fibration and the invariant splitting.

10. In both theorems, we assumed that $M \in C^{k,\alpha}_{b,u}$. Just as in the compact case, there is forced smoothness, see [HPS77]. That is, if we only have $M \in C^2_b$, while it is r-normally hyperbolic and the system is $C^{k,\alpha}_{b,u}$, then we have in fact $M \in C^{k,\alpha}_{b,u}$. Note that we do require $k \geq 2$ unlike the compact case, see also point 3.3.
This statement can be verified by reviewing the persistence proof under zero perturbation. We construct a smoothed approximation of M in Theorem 3.1 to model everything on, and in Theorem 3.2 we smoothen the linearization of the vertical part v_Y of the original vector field. These can be made as smooth as required, while we only use the perturbed flow (3.16) or the $C^{k,\alpha}_{b,u}$ smooth decomposition (3.24) (hence with remainder $\tilde{f} \in C^{k,\alpha}_{b,u}$) in the rest of the proof. Therefore the resulting manifold will satisfy $\tilde{M} = M \in C^{k,\alpha}_{b,u}$. ◊

These two theorems reduce to the corollaries formulated below, when M is compact or when $Q = \mathbb{R}^{m+n}$ with standard Euclidean metric and $M = \mathbb{R}^m \times \{0\}$. The statements then significantly reduce in complexity, and are comparable to well-known results.

Firstly, the case that M is compact. Then we can take a (pre)compact neighborhood $B(M;\varepsilon) \subset Q$ of M and thus conclude that bounded geometry holds on $\overline{B(M;\varepsilon)}$, ignoring irrelevant boundary problems. Any $C^{k,\alpha}$ function on $\overline{B(M;\varepsilon)}$ is automatically $C^{k,\alpha}_{b,u}$, so Theorem 3.1 reduces to the following corollary. For simplicity we leave out α-Hölder continuity and the C^k distance estimate between M and $\tilde{M}$.

Corollary 3.4 (Persistence of compact NHIMs) *Let $k \in \mathbb{Z}_{\geq 2}$, let (Q, g) be a smooth Riemannian manifold and $v \in C^k$ be a vector field on Q. Let $M \in C^k$ be a connected, compact submanifold of Q that is k-normally hyperbolic for the flow defined by v, with empty unstable bundle, i.e.* $\operatorname{rank}(E^+) = 0$.

Then for each sufficiently small $\eta > 0$ there exists a $\delta > 0$ such that for any vector field $\tilde{v} \in C^k$ with $\|\tilde{v} - v\|_1 < \delta$, there is a unique submanifold $\tilde{M}$ in the η-neighborhood of M, such that $\tilde{M}$ is diffeomorphic to M and invariant under the flow defined by $\tilde{v}$. Moreover, $\tilde{M}$ is C^k.

This corollary closely resembles [Fen72, Thm. 1] in the absence of a boundary ∂M (a boundary is allowed when M is overflowing invariant, see also Sects. 1.6.3 and 4.3). Note that our definition of normal hyperbolicity is less general (see Remark 1.10), and that we exclude unstable normal directions and the case[1] $k = 1$, while we do allow M to be an immersed submanifold. The persistence result of Hirsch, Pugh, and Shub [HPS77, Thm. 4.1 (f)] is similar to that in Fenichel's work; it additionally includes Hölder smoothness and allows immersed submanifolds as well.

Secondly, the case of $M = \mathbb{R}^m \times \{0\} \subset Q = \mathbb{R}^{m+n}$. Again, $\mathbb{R}^{m+n}$ has bounded geometry with one trivial, global chart. Thus, any object is $C^k_{b,u}$ if it can (locally) be described by $C^k_{b,u}$ functions, but with common global bound and continuity modulus. Then Theorem 3.2 reduces to the following corollary. We again suppress Hölder

[1] This was for technical reasons in the noncompact setting, see Remark 3.3, 5, and could be repaired in the compact setting.

continuity and drop the σ parameter dependence, which was only relevant for the reduction of Theorem 3.1.

Corollary 3.5 **(Persistence of a trivial NHIM in Euclidean space)** *Let $k \in \mathbb{Z}_{\geq 2}$. Let $v \in C^k_{b,u}$ be a vector field on $\mathbb{R}^{m+n}$ and let $M = \mathbb{R}^m \times \{0\}$ be a k-NHIM for the flow defined by v, with empty unstable normal bundle.*

Then for each sufficiently small $\eta > 0$ there exists a $\delta > 0$ such that for any vector field $\tilde{v} \in C^k_{b,u}$ with $\|\tilde{v} - v\|_1 < \delta$, there is a unique submanifold $\tilde{M} = \mathrm{Graph}(\tilde{h})$, $\tilde{h} : \mathbb{R}^m \to \mathbb{R}^n$, $\left\|\tilde{h}\right\|_0 \leq \eta$ such that $\tilde{M}$ is invariant under the flow defined by $\tilde{v}$. Moreover, $\tilde{h} \in C^k_{b,u}$ and $\left\|\tilde{h}\right\|_{k-1}$ can be made arbitrary small by choosing $\|\tilde{v} - v\|_k$ sufficiently small.

This theorem can be compared, for example, to [Sak90, Thm. 2.1]. Sakamoto's theorem is specifically targeted to singular perturbation problems. His conditions are more specific and concrete: the invariant manifold M is assumed to consist of stationary points and normal hyperbolicity is formulated in terms of the eigenvalues of normal derivatives of the vector field at M. He starts with an invariant manifold that is the graph of a nonzero function $h \in C^k_b$; this he reduces to the zero graph case $M = \mathbb{R}^m \times \{0\}$, while he incurs a loss of one degree of smoothness, obtaining a C^{k-1}_b persistent manifold and he requires $k \geq 3$, see [Sak90, p. 50]. He does allow both stable and unstable normal bundles.

In their series of papers [BLZ98; BLZ99; BLZ08], Bates, Lu, and Zeng obtained multiple results on noncompact NHIMs, including a persistence result. Most importantly, they work in Banach spaces with semi-flows, which adds nontrivial obstacles to be overcome, and allows application to PDE problems. On the other hand, my setting allows the ambient space to be a manifold, albeit finite-dimensional. They use the graph transform instead of the Perron method. This allows for the more general definition of relative normal hyperbolicity as in Remark 1.10. They include both stable and unstable normal directions, while they do not prove Hölder regularity. Finally, in [BLZ08] the interesting idea is developed to start with an approximate NHIM only.

If we ignore these differences, then their results fit in between the formulations of Theorem 3.1 and Corollary 3.5. Their invariant manifold M is immersed in a Banach space, but not necessarily described by the graph of a function h. Their hypothesis [BLZ08, p. 363] that the splitting does not twist too much is a bounded Lipschitz condition on the (approximate) splitting of the (un)stable and tangent bundles over M. This condition is similar, but slightly weaker than our condition $M \in C^2_{b,u}$, see also Remarks 3.3, 5 and 3.13.

Although the results of Bates, Lu, and Zeng are more general and complete in many aspects, I think that these cannot easily be generalized to prove a version of Theorem 3.1, set in an ambient manifold of bounded geometry. One could hope to use the Nash embedding theorem to obtain the ambient manifold (Q, g) as an isometrically embedded subspace of some $\mathbb{R}^n$. Then the dynamical system must be extended from Q to $\mathbb{R}^n$, such that M is still normally hyperbolic as a submanifold of $\mathbb{R}^n$; this procedure can be compared to the reduction in Sect. 2.6. The problem that arises is that the Nash embedding theorem provides no control on the extrinsic curvature of the embedding[2], so M need not be C_b^2 in $\mathbb{R}^n$. It might be possible to work around this by proving a 'bounded geometry version' of the Nash embedding theorem. Alternatively, their proof could likely be adapted to work with a uniform atlas of charts, e.g. using theory similar to that developed in Chap. 2.

We should also mention the paper [JS99] by Jones and Shkoller. They generalize Fenichel's results on persistence of overflowing invariant manifolds to semi-flows on infinite-dimensional Riemannian manifolds. They do assume the invariant manifold itself to be compact.

3.2 Outline of the Proof

The proof of Theorem 3.2 is lengthy and involves a lot of details. We therefore first present an overview of the separate steps involved in the proof.

First, in Sect. 3.4 we bring the system into a form that is suitable for application of further analytical techniques. That is, we decompose the vector fields along X and Y directions and linearize the vertical direction, leading to equations

$$\begin{aligned}\dot{x} &= v_X(x, y),\\ \dot{y} &= v_Y(x, y) = A(x)\, y + f(x, y)\end{aligned}$$

with a C^1-small term f. This allows us to apply a generalization of the classical Perron method for hyperbolic fixed points (see Sect. 1.4.2 for a quick overview) to NHIMs, first presented by Henry [Hen81, Chap. 9]. In the case of a hyperbolic fixed point, we could fully linearize the system; here, we can only linearize the normal directions, while we keep the full nonlinear form in the directions along X. These cannot be linearized because we have no control to localize the dynamics in the directions along X.

In the theorem, the invariant manifold M is given as a (small) graph $h : X \to Y$. A coordinate change to represent the invariant manifold as $M = X \times \{0\}$ would (re)introduce a loss of smoothness that we carefully worked around in Chap. 2 by means of a uniformly smoothed submanifold. The graph h can be chosen arbitrarily close to the zero section, and together with the small perturbation $\tilde{v} - v$, this influences the exponential growth rates (1.10) only slightly. We recover Eqs. (3.16) for the perturbed system that satisfy slightly perturbed exponential estimates (3.17), even when we decouple the equations for x and y by inserting curves $y(t)$ and $x(t)$,

---

[2] This can be seen from the result that the Nash embedding can be obtained into an arbitrarily small ball. As an explicit example, take $Q = \mathbb{R}$ with standard metric and embed it into $\mathbb{R}^2$ via the map $r(\theta) = \arctan(\theta)/\pi + \frac{1}{2}$ in polar coordinates. Since the integral of $r(\theta)$ diverges both when $\theta \to \pm\infty$, we obtain (after arc length reparametrization) an isometric embedding of $\mathbb{R}$ 'curled up' into $B(0; 1)$, while the extrinsic curvature grows unbounded for $\theta \to -\infty$.

respectively, that are 'close' to solution curves of the original system. We directly include the perturbation $\tilde{v} - v$ into the horizontal component of the vector field; for the vertical component we include the perturbation in the nonlinear term $\tilde{f}$. This gives rise to a nonlinear, horizontal flow $\Phi_y(t, t_0, x_0)$ and a linear, vertical flow $\Psi_x(t, t_0)\, y_0$ that depend on a curve in the other space and satisfy estimates

$$\forall t \leq t_0 : \left\| \mathrm{D}\Phi_y(t, t_0, x_0) \right\| \leq C_X\, e^{\rho_X\, (t - t_0)},$$
$$\forall t \geq t_0 : \| \Psi_x(t, t_0) \| \leq C_Y\, e^{\rho_Y\, (t - t_0)},$$

where ρ_X, ρ_Y are close to the original exponential rates $-\rho_M$, ρ_-.

The next step in Sect. 3.6 is to define a pair of maps (3.32) between curves in X and Y in terms of these flows and the decomposed vector fields (3.16). The composition $T = T_Y \circ (T_X, \mathrm{pr}_1)$ of these maps will be a contraction on bounded curves in Y depending on a parameter $x_0 \in X$, but we measure these curves with $\|\cdot\|_\rho$ norms for some exponent ρ with $\rho_Y < \rho < \rho_X$. Lemma 3.27 shows that the fixed points of T are precisely the vertical parts $y(t)$ of solution curves of the perturbed vector field $\tilde{v}$ that stay in the tubular neighborhood $\|y\| \leq \eta$ of M and have initial value x_0 for their horizontal part $x(t)$. The maps T_X and T_Y generalize the center-unstable and stable components, respectively, of the Perron integral in the fixed point case, cf. Sect. 1.4.2. The nonlinear flow along the invariant manifold is used in T_X, but now depends on the vertical component $y(t)$ too, while in the vertical, normal directions we use a variation of constants integral to separate the nonlinear terms from the linearized flow, just as in the classical Perron method.

This setup leads to a fixed point map $\Theta : X \to B^\rho_\eta(I; Y)$ that maps an initial value $x_0 \in X$ to the unique bounded curve in Y that corresponds to a full solution curve (x, y) such that $x(0) = x_0$. If we now evaluate the vertical solution curve $y = \Theta(x_0)$ at $t = 0$, then we obtain the vertical component $y_0 = y(0) \in Y$ of the initial value corresponding to $x_0 \in X$. All these solution curves stay close to M and form an invariant manifold, so the graph of

$$\tilde{h} : X \to Y : x_0 \mapsto \Theta(x_0)(0)$$

must describe the unique perturbed invariant manifold $\tilde{M}$. Application of the contraction principle immediately implies that $\tilde{h}$, and therefore $\tilde{M}$, is Lipschitz continuous.

In Sect. 3.7 we continue to prove that $\tilde{M}$ is $C^{k,\alpha}$. We start that section with a more detailed overview of this smoothness part of the proof and in Sect. 3.7.1 we present a scheme to obtain the first derivative in a number of steps. Higher smoothness then follows along the same lines, just with more complex expressions, see Sect. 3.7.9. Let us focus here on the basic ideas.

Smoothness of M follows directly from smoothness of Θ. We study the derivatives of Θ by formal differentiation of the fixed point equation; this leads to Eq. (3.42). But let us consider for a moment a simpler heuristic formulation, similar to Eq. (1.8) for the hyperbolic fixed point in Sect. 1.4.3. Then the derivatives of the Perron fixed point map are

$$\mathrm{D}^k T(y)\big(\delta y_1, \ldots, \delta y_k\big)(t) = \int_{-\infty}^{t} \Psi(t,\tau) \cdot \mathrm{D}^k \tilde{f}(y(\tau))\big(\delta y_1(\tau), \ldots, \delta y_k(\tau)\big) \mathrm{d}\tau.$$

Even if $\mathrm{D}^k \tilde{f}$ is bounded, it acts as a multilinear map on a k-tuple of variations δy_i, each having exponential growth of order ρ, so the result has exponential growth of order $k\,\rho$. This is canceled by the exponential growth of $\Psi(t,\tau)$ if $\rho_Y < k\rho$. Then $\mathrm{D}^k T$ can be viewed as a contraction on $\mathrm{D}^k\Theta$, but only when $\mathrm{D}^k\Theta$ is viewed as a map into $B^{k\,\rho}(I; Y)$. To obtain continuity of the maps $\mathrm{D}^k T$, we have to add another arbitrarily small term $\mu < 0$ to the exponent, i.e. $k\,\rho + \mu$; in case of α-Hölder continuity we need $\mu = \alpha\,\rho$. These key facts show how the spectral gap condition limits smoothness; see Sect. 1.2.1 for a detailed discussion and an example that shows that our smoothness result is in fact sharp.

The technique of using a scale of Banach spaces as developed by Vanderbauwhede and Van Gils [VG87], and the fiber contraction theorem of Hirsch and Pugh [HP70] can be applied, and we obtain each $\mathrm{D}^k\Theta$ as a fixed point in the appropriate space. The final conclusion $\tilde{h} \in C^k$ follows by evaluating $\mathrm{D}^k\Theta$ at $t = 0$.

We have to be very careful however: higher derivatives of maps between manifolds are difficult to define (at least in a practical way), so we develop some theory to describe higher derivatives using normal coordinates in Appendix C and generalize results from $\mathbb{R}^n$ to this setting. Secondly, the derivatives of T_X, T_Y only exist as 'formal derivatives' on 'formal tangent bundles'. We endow these formal tangent bundles with a topology induced by parallel transport. This allows us to study continuity of the formal derivatives $\underset{\sim}{\mathrm{D}} T_X$, $\underset{\sim}{\mathrm{D}} T_Y$ at the cost of introducing additional holonomy terms. Finally, we do obtain the $\mathrm{D}^k\Theta$ maps as derivatives of Θ.

3.3 Compactness and Uniformity

The classical results on normally hyperbolic invariant manifolds [Fen72; HPS77] assume the invariant manifold to be compact. This is used to obtain uniform boundedness and continuity of the vector field and other objects. Here, instead, we assume these objects to have the required uniformity directly, replacing the compactness requirement. In this section, we expose some of the issues that need to be dealt with and we present accompanying examples. We focus here on those issues that are not (clearly) present in the literature and only show up when considering general manifolds. See Sect. 1.2.2 for motivation and examples of noncompact NHIMs.

The primary requirement in the noncompact case—well-known to experts in the field—is that the vector field defining the system must be uniformly bounded, including all its spatial derivatives up to the order of the smoothness result requested. Secondly, the vector field should be uniformly continuous, and uniformly α-Hölder continuous when $\alpha \neq 0$. The case $\alpha = 0$ really is a special case, whose proof needs more care. Hölder continuity provides an explicit continuity estimate, which is tailored to the problem; 'plain' uniform continuity does not provide this, forcing us to use an (arbitrarily) small amount of the spectral gap to compensate.

In the compact case, Fenichel [Fen72, p. 200] argues that persistence of the invariant manifold should be independent of the choice of a Riemannian metric. Indeed, he proves that the exponential growth rates are independent of such a choice, as all metrics are equivalent. In a noncompact setting, however, non-equivalent metrics do exist and we do expect persistence to depend on the choice of metric, since it determines which perturbations are globally C^1 small. Moreover, we make a technical uniformity assumption of *bounded geometry* (see Chap. 2, also for example spaces of bounded geometry) on both the underlying space and the invariant manifold. These assumptions are automatically satisfied in the compact case. It is not clear though to what extend they are essential in the noncompact case.

The remainder of this section is devoted to examples that show multiple aspects that should be treated carefully in the noncompact setting, while being trivially fulfilled in the compact case. Some interesting examples can also be found in the early work [Hop66] by Hoppensteadt. He presents counterexamples to uniform stability of solutions in a time-dependent singular perturbation setting when the stability criteria do not have sufficient uniformity.

3.3.1 Non-Equivalent Metrics

As a simple example of two metrics leading to different results in the noncompact case, let us consider the following.

Example 3.6 (*Non-equivalent metrics*) Let $X = \mathbb{R}^2$, $Y = \mathbb{R}$ with on the one hand the usual Euclidean metric g_e and on the other hand a metric g_s induced by a diffeomorphism similar to stereographic projection from the sphere (with North Pole removed) onto $X = \mathbb{R}^2$.

Let the vector field be given by

$$v(x, y) = \big(\arctan(\|x\|)\,\frac{x}{\|x\|},\ \lambda\, y\big) \tag{3.1}$$

with $\lambda < 0$. This makes the vertical, y direction uniformly attracting with exponent λ, while in the plane $X \times \{0\}$ the origin is an expanding fixed point and the exponential growth rate is everywhere non-negative. Thus, $M = X \times \{0\}$ is an r-NHIM for arbitrarily large r and $v \in C^k_{b,u}$ for any $k \geq 1$ on a tubular neighborhood of M of size $\eta = 1$, say. See Fig. 3.1 on the left side.

In polar coordinates (s, θ) on X we have

$$v(s, \theta, y) = \big(\arctan(s),\ 0,\ \lambda\, y\big).$$

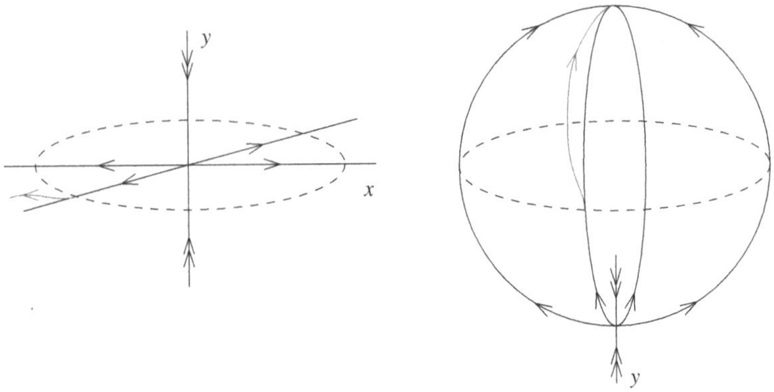

Fig. 3.1 A normally hyperbolic flow with respect to Euclidean and sphere induced metrics

Now, instead of the usual stereographic projection map $(\varphi, \theta) \mapsto (s = \tan(\varphi/2), \theta)$ with $\varphi = 0$ at the South Pole, we take

$$f : [0, \pi) \to \mathbb{R}_{\geq 0} : \varphi \mapsto -\log(1 - \varphi/\pi) \tag{3.2}$$

and the corresponding diffeomorphism Φ that acts trivially along the directions of θ and y coordinates. This diffeomorphism induces a metric g_s on X by pushforward of the standard metric on the sphere. The system with metric g_s is most easily studied by pullback to the sphere with North Pole removed, see Fig. 3.1 on the right side. This is an equivalent formulation since Φ is an isometry by construction. The vector field is then represented on $S^2 \times \mathbb{R}$ by

$$(\Phi^* v)(\varphi, \theta, y) = \big(\pi(1 - \varphi/\pi) \arctan\big(-\log(1 - \varphi/\pi)\big),\ 0,\ \lambda\, y\big). \tag{3.3}$$

This shows that the vector field is still C_b^1, although not $C_{b,u}^1$ anymore in this metric. More importantly, the system can be extended to include the North Pole as an attracting fixed point. The rate of attraction along the perpendicular y direction has not changed from λ, but along the horizontal directions of the sphere, the attraction rate now is

$$\lim_{\varphi \to \pi} \mathrm{D}_1(\Phi^* v)(\varphi, 0, 0) = -\frac{\pi}{2}. \tag{3.4}$$

In both metrics the normal exponential attraction rate is $\rho_Y = \lambda < 0$ since the linear flow on Y is decoupled from X. The exponential growth rate on X does depend on the choice of metric. For the Euclidean metric g_e, we have $\rho_X = 0$. This follows from an analysis of the radial component $\dot{s} = \arctan(s)$ of the system. At $s = 0$ this has unstable exponent 1, while away from $s = 0$ the tangent flow is uniformly bounded away from both zero and infinity, so there the Lyapunov exponent is zero. With respect to the metric g_s, it follows from 3.4 that the Lyapunov exponent is $\rho_X = -\frac{\pi}{2}$ for solutions approaching planar infinity. Thus, we see that if $\lambda \geq -\frac{\pi}{2}$, then the system is not normally hyperbolic with respect to the metric g_s, while

$X \times \{0\}$ is an r-NHIM for any $r \geq 1$ under the metric g_e. On the other hand, if $\lambda < -\frac{\pi}{2}$, then the system is still only r-normally hyperbolic with respect to g_s for $r < \pi/(2\lambda)$. Again, we can construct an explicit perturbation similar to that in Example 1.1. If we add a small vertical perturbation $\varepsilon\,\chi$ with support away from the poles (compare with Fig. 1.3), then along meridians passing through supp χ, the invariant manifold is lifted and approaches the North Pole $\varphi = \pi$ approximately along a graph $y = C\,(\pi - \varphi)^{-2\lambda/\pi}$, while on meridians not passing through supp χ, the invariant manifold stays at $y = 0$. See the perturbed flow lines in Fig. 3.1. This results in unbounded C^r derivatives of the perturbed invariant manifold at the North Pole for $r > -2\lambda/\pi$. ○

We conclude that in the noncompact setting, normal hyperbolicity explicitly depends on the choice of metric since metrics need not be equivalent. Moreover, the allowed size of the perturbations depends on the metric.

Example 3.7 (*Perturbation sizes depend on the choice of metric*) We extend Example 3.6 above. Let $\lambda < -\frac{\pi}{2}$ so that the system is normally hyperbolic with respect to both metrics, and set

$$w(s, \theta, y) = \big(0,\ 0,\ \arctan(s)\ \sin(\theta)\big). \tag{3.5}$$

Then the vector field $v + \varepsilon\, w$ is a C^1 small perturbation of 3.1 with respect to g_e and perturbs the original manifold M smoothly to a manifold $\tilde{M}$ that has height converging to $y = -\varepsilon\ \sin(2\,\theta)\,\pi/(2\lambda)$ along radials when $s \to \infty$. The pullback of $\tilde{M}$ to the sphere, however, has a discontinuity at the North Pole, since $s \to \infty$ corresponds to $\varphi \to \pi$, and so the North Pole is approached at different constant heights along these radials. This apparent contradiction that $\Phi^*(\tilde{M})$ is not a C^1 small perturbation with respect to g_s stems from the fact that w is not C^1 small with respect to this metric. The vector field w has unbounded derivatives since g_s 'squeezes' distances when approaching planar infinity, that is, the North Pole. ○

Thus, non-equivalent metrics also lead to different classes of C^1 small perturbations under which the invariant manifold persists. Moreover we see that one cannot simply get rid of noncompactness by a compactification argument. The metric g_s is induced by a one-point compactification to the sphere, but leads to different normal hyperbolicity properties than the noncompact case with metric g_e. Any other choice of the diffeomorphism Φ would lead to the same problems, since pullback of the metric g_e must introduce a singularity at the North Pole. This cannot be equivalent to a metric that extends regularly there.

3.3.2 Non-Persistence of Embedded NHIMs

Let us give another example which shows that an embedded invariant manifold need not persist. This example clarifies the remarks already made in the introduction in

Sect. 1.6.2: noncompact embedded NHIMs can perturb into immersed manifolds. On the one hand, this example shows that it is natural to consider immersed NHIMs. It also shows that a noncompact NHIM must have a uniformly sized tubular neighborhood that does not self-intersect, in order to guarantee perturbation as an embedded manifold. Further details can be found in Sect. 2.3 where the concept of a uniformly embedded submanifold is defined.

In the example presented here, the unperturbed manifold is normally hyperbolic but noncompact and 'touches' itself in the limit to infinity, see Fig. 3.2, the top image. In this case, we can find arbitrarily small perturbations that will let the two persisting branches collapse into one at a finite point.

Example 3.8 (A non-uniformly embedded NHIM) Let $(x, y) \in \mathbb{R}^2 = Q$. For $x \leq 0$ we define the vector field v of the system in polar coordinates, and for $x \geq \frac{1}{2}$ in Cartesian coordinates as

$$\begin{array}{ll} (\dot{r}, \dot{\theta}) = v(r, \theta) = (1 - r\,,\, -\sin(\theta/2)) & \text{if } \; x = r\,\cos(\theta) \leq 0, \\ (\dot{x}, \dot{y}) = v(x, y) = (e^{-x}\,,\, -y)\text{if} & \qquad x \geq \frac{1}{2}. \end{array} \tag{3.6}$$

We glue these vector fields together in a smooth way somewhere between $x = 0$ and $x = \frac{1}{2}$. Then the manifold as shown at the top in Fig. 3.2 is a NHIM. The flow attracts uniformly in the normal direction with rate -1 (except that the rate may deviate slightly around the glued area), while along the manifold, the flow has an expanding fixed point at $(-1, 0)$ and the contraction in the direction of $x \to \infty$ is weaker than exponential. Explicitly solving the flow for $x \geq \frac{1}{2}$ yields

$$\Phi^t(x, y) = (\log(e^x + t)\,,\, y\,e^{-t}),$$

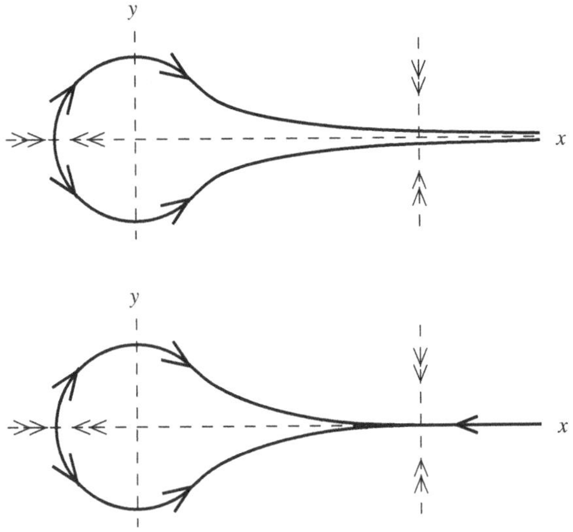

Fig. 3.2 Collapse of a nearly self-intersecting invariant manifold

which exhibits the rates of contraction in the normal and tangential directions by considering either projection in

$$\lim_{t\to\infty} \frac{1}{t} \log\left(\pi_{x,y} \circ \mathrm{D}\Phi^t(x, y)\right).$$

Let us now introduce the very simple perturbation vector field $w(x, y) = (-\varepsilon\,, 0)$ for $x \geq 1$ and smoothly cut off to zero left of $x = 1$. When this perturbation is added to the vector field v, the vertical line $x = -\log(\varepsilon)$ becomes a stable, invariant set, see Fig. 3.2 the bottom image. The upper and lower branch of the original NHIM will both converge to the newly created fixed point $(-\log(\varepsilon)\,, 0)$. On the right side of this point the manifold is given by the single line $y = 0$.

Each branch separately persists as a C^∞ manifold, as could (naively) be expected. The problem is that we have no control on the distance between the two branches, so for any $\varepsilon > 0$, these branches will collapse at some point where the persisting object ceases to be an embedded manifold. As already remarked, there are two ways to address this issue. One can abandon the implicit assumption that the NHIM is an embedded submanifold and replace this by immersed submanifolds; this idea was introduced already in [HPS77]. If one insists on having embedded submanifolds, even under perturbations, then one must eliminate the possibility of these 'collapses' occurring. A sufficient condition is the existence of a uniformly sized tubular neighborhood of the invariant manifold that does not intersect itself. Global control on the perturbation distance of the invariant manifold will imply that the perturbed manifold stays inside this tubular neighborhood and thus will not self-intersect. ○

3.3.3 *Non-Uniform Geometry of the Ambient Space*

The previous examples were set in Euclidean space. The next two examples show that additional uniformity conditions must be imposed on a nontrivial ambient space. It is not enough to assume uniform continuity and boundedness for the dynamical system. The first example is an extension to the previous one and shows that the ambient space must have a uniformly finite injectivity radius. The second example indicates that even if the ambient space has finite injectivity radius and trivial topology, persistence might be lost due to non-bounded curvature of the ambient space.

Example 3.9 (*Zero* injectivity radius) We construct as ambient space Q a cylinder whose radius shrinks exponentially. That is, we take $Q = \mathbb{R} \times S^1$ with metric $g(x, \theta) = \mathrm{d}x^2 + e^{-2x}\mathrm{d}\theta^2$. See Fig. 3.3 for an impression, but note that the metric induced by the embedding in $\mathbb{R}^3$ is not (and cannot be made) the same as g. The vector field

$$v(x, \theta) = (1, 0) \tag{3.7}$$

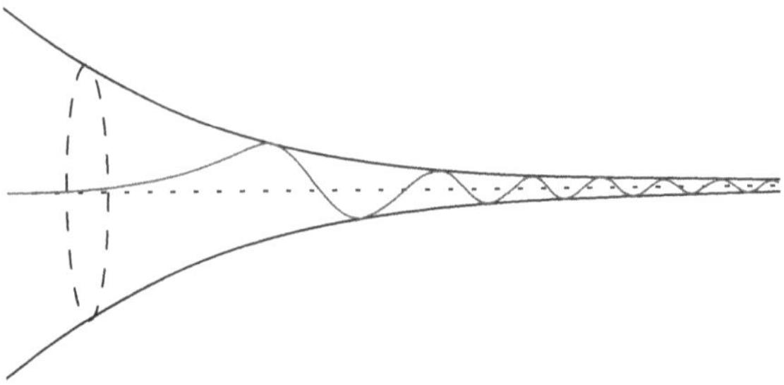

Fig. 3.3 A non-uniform cylinder with a winding curve

generates a simple flow along the cylinder, and *each* solution curve is a NHIM purely due to the fact that all curves flow into an exponentially shrinking tube, while there is no contraction along the curve.

Let us consider the invariant manifold $M = \{\theta = 0\}$. We add a perturbation to the vector field that is given by $\dot{\theta} = \varepsilon$ for $x \geq 0$ and is smoothly cut off to zero left of $x = 0$. This perturbation is smooth and C^1 small with respect to the metric[3] g. When the original curve M enters the region $x \geq 0$, it is modified to a curve $\tilde{M}$ that starts winding around the cylinder, as indicated in Fig. 3.3.

This clearly cannot be represented in a tubular neighborhood of M in Q since the curve would leave the neighborhood 'above' and reenter 'from below'. On the other hand, the normal bundle of M can be viewed as a covering of Q, and on that covering, $\tilde{M}$ is represented by the function $\theta(x) = \varepsilon\, x$, which is still a bounded graph with norm $\|\theta(x)\| = \varepsilon\, x\, e^{-x}$, but which winds around since $\theta \in [0, 2\pi)$. Thus, a globally finite injectivity radius seems a necessary requirement if we want the perturbed manifold to be represented in a diffeomorphic tubular neighborhood of M. ○

The second example indicates that a finite injectivity radius is not enough; unbounded curvature of the ambient manifold might lead to loss of persistence of the NHIM. It should be pointed out that this example satisfies all properties of normal hyperbolicity with uniform estimates up to C^1 smoothness, except that the vector field has no uniformly continuous derivative. I have not been able to add this final property to create a complete counterexample where persistence fails in the absence of the curvature property of bounded geometry only.

Example 3.10 (Unbounded curvature) Let $Q = \mathbb{R}^3$ with metric

$$g(x, y, z) = \mathrm{d}x^2 + \exp\big(-2\,|x|\arctan(x\,z)\big)\,\mathrm{d}y^2 + \exp\big(-2\,|x|\big)\,\mathrm{d}z^2, \qquad (3.8)$$

but with component functions symmetrically smoothed around $x = 0$. This Riemannian manifold is invariant under translations in y and has a mirror symmetry involution in any plane of fixed y. Hence, each submanifold $\{\, y = y_0 \,\}$ is geodesically invariant.

[3] Measuring the C^1 size with respect to g requires taking covariant derivatives and may introduce results not directly apparent in coordinates (x, θ). A perturbation term $\dot{\theta} = \varepsilon\, \exp(x)$ would still be globally small in this metric.

Let the vector field be

$$v(x, y, z) = \begin{cases} \big(x,\ 0,\ -\arctan(z)\big) & \text{for } |x| \le 1, \\ \big(\operatorname{sign}(x),\ 0,\ 0\big) & \text{for } |x| \ge 1, \end{cases} \tag{3.9}$$

and smoothly glued together in a neighborhood of the boundary $|x| = 1$, see Fig. 3.4. Thus, the whole system is invariant under translations in y, and within any plane $\{ y = y_0 \}$, the point $(x, z) = (0, 0)$ is a hyperbolic fixed point with eigenvalues 1 and -1 in the x and z direction, respectively. The system also has a mirror symmetry around $x = 0$; from now on we only consider $x \ge 0$.

The plane $M = \{ z = 0 \}$ is a NHIM; it is clearly invariant under the flow, and similar to the exponentially shrinking cylinder, the metric contracts in the z direction along solution curves $x \to \infty$, while no contraction occurs along the manifold. On $\mathrm{T}M$ the metric reduces to $g|_{\mathrm{T}M} = \mathrm{d}x^2 + \mathrm{d}y^2$ while the flow is linear in time. On a neighborhood of the y-axis, finally, normal hyperbolicity follows from the attraction along the z directions due to the term $-\arctan(z)$ in (3.9).

The vector field v and its covariant derivative are uniformly bounded with respect to the metric. For $x \le 2$, this follows from the fact that g including its inverse and derivatives, as well as v and its derivatives are bounded. For $x \ge 2$, explicit calculations in local coordinates show that $\|v\| = 1$, while we have

$$\nabla v = \begin{pmatrix} 0 & 0 & 0 \\ 0 & -\frac{x\,z}{1+(x\,z)^2} - \arctan(x\,z) & 0 \\ 0 & 0 & -1 \end{pmatrix},$$

expressed in an orthonormal frame, which is bounded as well. The second covariant derivative $\nabla^2 v$ is unbounded, though. This indicates that ∇v is probably not uniformly continuous for a reasonable definition of uniform continuity, cf. Definition 2.9, although I have not completely investigated this question.

The Ricci scalar curvature of Q is unbounded and on $M = \{ z = 0 \}$ it is given by

$$S = -2\left(1 + x^4\, e^{2x}\right).$$

Clearly, this implies that the Riemannian curvature is unbounded too.

Remark 3.11 In hindsight, it should probably not come as a complete surprise that $\nabla^2 v$ is unbounded. The Riemannian curvature is unbounded, and since it is the generator of holonomy (see Sect. 2.2), it can thus generally be expected that holonomies along infinitesimal loops act as an unbounded family of operators on v. These are expressed in local coordinates by second covariant derivatives of v:

$$R(\partial_i, \partial_j)\, v = \nabla_{\partial_i} \nabla_{\partial_j} v - \nabla_{\partial_j} \nabla_{\partial_i} v. \qquad \lozenge$$

We proceed with checking that $\exp : N \to Q$ has finite injectivity radius on the normal bundle[4] N of M. Then all assumptions for persistence are fulfilled, except for bounded curvature (and uniform continuity of ∇v). As each submanifold $\{ y = y_0 \}$ is invariant, we can restrict our investigation to $y = 0$, such that x denotes the coordinate along the base manifold of the normal bundle; let t denote the normalized coordinate in the vertical direction. The exponential map of (x, t) is generated by the geodesic flow as follows: start at x with vertical unit vector and then follow a geodesic for time t. For $x = 2$, say, this flow is well-defined and stays inside the region $x \geq 1$ for some bounded time $|t| \leq r$. The diffeomorphism group

$$\varphi(x, z) = (x + \xi, z\, e^{\xi}) \qquad \text{with} \quad \xi \in \mathbb{R}$$

translates along x while simultaneously scaling z, see also Remark 2.2. In the region $x \geq 1$ this is an isometry, so the exponential mapping defined for $x = 2,\ |t| \leq r$ can be isometrically mapped onto the whole region $x \geq 2,\ |t| \leq r$. For x on the compact interval $[0, 2]$ the exponential map must have a finite injectivity radius too, so there exists a global $r > 0$ such that $\exp : N_{\leq r} \to Q$ is diffeomorphic onto its image.

Now we add a perturbation in a similar spirit to that in Sect. 1.2.1: we lift M by a local, vertical perturbation of the vector field, varying along y. In a neighborhood of the plane $x = 2$ we add a small vertical component

$$\dot{z} = \varepsilon\,(2 - \cos(y))\, \exp\Big(\frac{-1}{1 - (x-2)^2}\Big) \qquad \text{if} \quad |x - 2| \leq 1.$$

In the region $x \leq 1$ the flow is unmodified, so there the perturbed manifold $\tilde{M}$ must coincide with the original M; otherwise it would not stay in a bounded neighborhood of M under the backward flow. Around $x = 2$, the flow lifts $\tilde{M}$ to at least a height

$$z \geq \varepsilon \int_1^3 \exp\Big(\frac{-1}{1 - (x-2)^2}\Big) \mathrm{d}x \geq \varepsilon/4$$

and the height z depends on y, see Fig. 3.5. Then in the region $x \geq 3$ the manifold $\tilde{M}$ continues along $\dot{x} = 1$ at the same y, z coordinates. Now we have $\varepsilon/4 \leq z \leq 6\,\varepsilon$, so for all small $\varepsilon > 0$ the y component of the metric along the flow on the invariant manifold eventually shrinks at an exponential rate that is stronger than in the z direction, while $\tilde{M}$ has variable height $z(y)$ independent of x. Hence the Lipschitz norm of $\tilde{M}$ can be estimated by measuring $z'(y)$ with respect to the metric g along the manifold. But $z'(y)$ is nonzero and constant along x in coordinates, while horizontal distances along y shrink faster than vertical distances. This means that the Lipschitz norm of $\tilde{M}$ grows unbounded for $x \to \infty$. Moreover, the normal exponential growth

[4] We should actually show that the injectivity radius of Q is finite, i.e. $r_{\mathrm{inj}}(Q) > 0$, at least in a neighborhood of M. I have not been able to do this. Finite injectivity radius of the normal bundle does allow us to construct a tubular neighborhood to model persistent manifolds close to M, though.

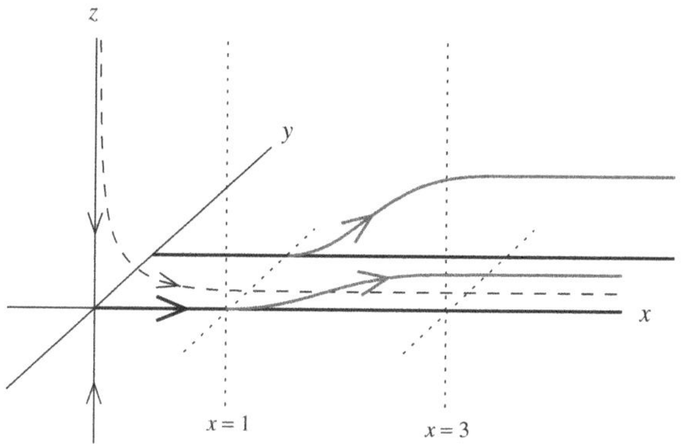

Fig. 3.4 Perturbation of a normally hyperbolic system in unbounded geometry

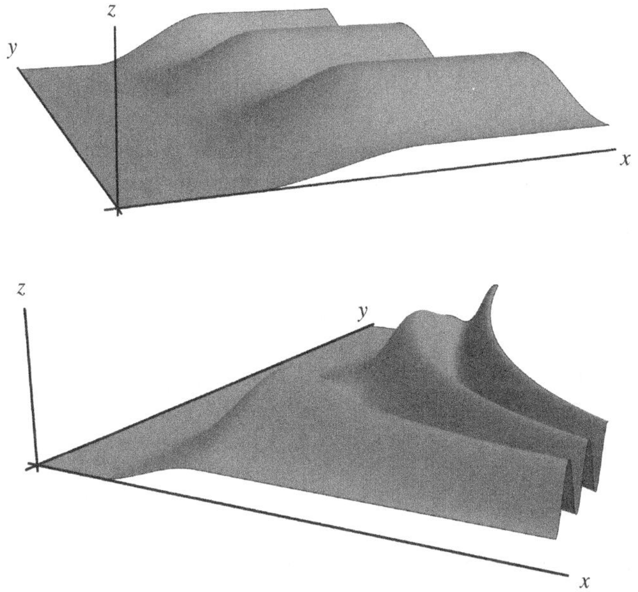

Fig. 3.5 The graph of the perturbed manifold with respect to the Euclidean metric (*top*) and an approximate image of the same graph with respect to the metric g (*bottom*)

rate does not dominate the tangential rate anymore, so the perturbed manifold is not normally hyperbolic anymore. ○

3.4 Preparation of the System

As a first step towards proving Theorem 3.2 we shall bring the system in a form suitable to apply analytical tools to it. Let

$$\begin{aligned} \dot{x} &= v_X(x, y) \in \mathrm{T}_x X, \\ \dot{y} &= v_Y(x, y) \in Y, \end{aligned} \tag{3.10}$$

be the decomposition of the vector field v along X and Y. The invariant manifold is given as the graph $M = \{y = h(x)\}$. We dropped the explicit dependence on σ from the notation. The full flow of v will be denoted by Υ^t, while Φ, Ψ are reserved for flows defined in terms of the horizontal and vertical components of v, respectively. To shorten notation we write $g(x) = (x, h(x))$. We shall always assume that $\|y\| \le \eta$.

Our goal is to establish a linearized form

$$v_Y(x, y) = A(x)\, y + f(x, y) \tag{3.11}$$

for the vertical part of (3.10) such that f is small and $A, f \in C^{k,\alpha}_{b,u}$, while the flows Φ^t and Ψ^t generated by

$$\begin{aligned} \dot{x} &= v_X(g(x)), \\ \dot{y} &= A(x(t))\, y \qquad \text{with } g(x(t)) \text{ a solution curve on } M, \end{aligned} \tag{3.12}$$

should satisfy exponential growth estimates (1.10) as in Definition 1.8 of normal hyperbolicity with exponents ρ_X, ρ_Y close to the original $-\rho_M$, ρ_-, respectively; the corresponding constants $\tilde{C}_M$, $\tilde{C}_-$ may differ arbitrarily from the original C_M, C_-.

We first identify the invariant splitting and associated flows on $\mathrm{T}_M(X \times Y)$ to be able to relate these exponential growth rates, see also Fig. 3.6. By definition of normal hyperbolicity (without an unstable bundle) we have

$$\mathrm{T}_M(X \times Y) = \mathrm{T}M \oplus E^-, \quad \mathbb{1} = \pi_{\mathrm{T}M} + \pi_{E^-}, \quad \mathrm{D}\Upsilon^t = \mathrm{D}\Upsilon^t_M \oplus \mathrm{D}\Upsilon^t_-$$

with associated exponential growth rates (1.10). On the other hand we have the splitting

$$\mathrm{T}(X \times Y) = \pi_Y^*(\mathrm{T}X) \oplus \pi_X^*(\mathrm{T}Y) \cong \mathrm{T}X \times (Y \times Y)$$

that is naturally induced by the trivial bundle structure. The identification $\mathrm{D}g = \mathbb{1}_{\mathrm{T}X} + \mathrm{D}h : \mathrm{T}X \to \mathrm{T}M$ is bounded linear with bounded inverse, so the associated

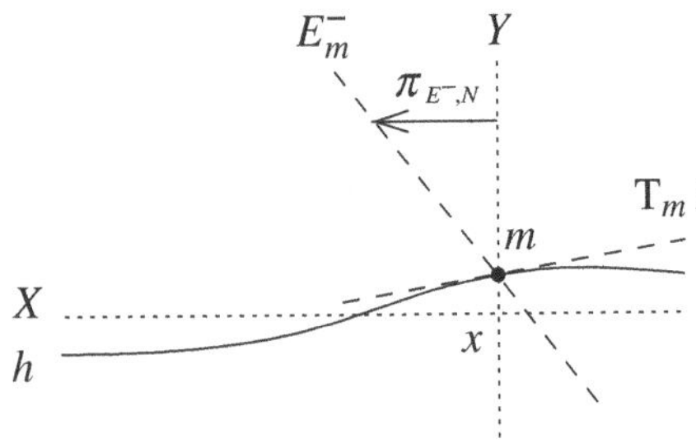

Fig. 3.6 The splitting $\mathrm{T}_m(X \times Y) = \mathrm{T}_m M \oplus E^-_m$ with $m = g(x)$

vector field $g^*(v) = v_X \circ g$ on X generates a flow Φ^t such that $\mathrm{D}\Phi^t$ has the same exponential growth rate as $\mathrm{D}\Upsilon^t_M$, up to a bounded factor $\|\mathrm{D}g^{-1}\| \cdot \|\mathrm{D}g\|$ due to the norms on the different tangent spaces. Recall that v_σ depends on a parameter $\sigma \in (0, \sigma_0]$. We choose the bound σ_1 small enough such that for all $\sigma \leq \sigma_1$ we have

$$\forall\, t \leq 0: \|\mathrm{D}\Phi^t\| \leq \tilde{C}_M\, e^{-\rho_M t} \quad \text{with} \quad \tilde{C}_M = 2\, C_M.$$

Let $N = \pi_X^*(\mathrm{T}Y)|_M$ denote the vertical bundle over M, whose fibers can be canonically identified with Y. Just as above, we want to project the flow $\mathrm{D}\Upsilon^t_-$ onto N while preserving the exponential growth rate. The projection π_{E^-} along $\mathrm{T}M$ is uniformly bounded for all σ. This means that the angle between $\mathrm{T}M$ and E^- is bounded away from zero. Since $\mathrm{T}M$ can be chosen arbitrarily close to the horizontal $\mathrm{T}X$ by choosing σ sufficiently small, it follows that the projection $\mathrm{D}\pi_Y|_{E^-} : E^- \to N$ and its inverse $\pi_{E^-,N}$ are bounded for all $\sigma \leq \sigma_1$ when σ_1 is sufficiently small, see also Fig. 3.6. To this end, let $(0, \varphi) \in \mathrm{T}_{g(x)}(X \times Y)$ and consider the identity

$$\varphi = \mathrm{D}\pi_Y \cdot (0, \varphi) = \mathrm{D}\pi_Y \cdot (\pi_{\mathrm{T}M} + \pi_{E^-}) \cdot (0, \varphi).$$

We have $\pi_{\mathrm{T}M} \cdot (0, \varphi) \in \mathrm{T}M$ so $\pi_{\mathrm{T}M} \cdot (0, \varphi) = (\xi, \mathrm{D}h(x)\,\xi)$ where $\xi = \mathrm{D}\pi_X \cdot \pi_{\mathrm{T}M} \cdot (0, \varphi) \in \mathrm{T}_x X$ Now we have estimates

$$\begin{aligned} \|\xi\| &= \|\mathrm{D}\pi_X\, \pi_{\mathrm{T}M}\, (0, \varphi)\| \leq \|\pi_{\mathrm{T}M}\|\, \|\varphi\|, \\ \|\pi_{\mathrm{T}M} \cdot (0, \varphi)\| &= \|\mathrm{D}h(x)\,\xi\| \leq \sigma\, \|\pi_{\mathrm{T}M}\|\, \|\varphi\|, \\ \|\mathrm{D}\pi_Y \cdot \pi_{E^-} \cdot (0, \varphi)\| &= \|\varphi - \mathrm{D}\pi_Y \cdot \pi_{\mathrm{T}M} \cdot (0, \varphi)\| \geq (1 - \sigma\, \|\pi_{\mathrm{T}M}\|)\, \|\varphi\| \end{aligned}$$

from which it follows that $\mathrm{D}\pi_Y|_{E^-}$ has an inverse $\pi_{E^-,N} : N \to E^-$ for which we have the bound $\|\pi_{E^-,N}\| \leq (1 - \sigma_1\, \|\pi_{\mathrm{T}M}\|)^{-1}\, \|\pi_{E^-}\| \leq 2\, \|\pi_{E^-}\|$ if we choose $\sigma_1 \leq \frac{1}{2\|\pi_{\mathrm{T}M}\|}$.

Consider the flow

$$\hat{\Psi}^t = \mathrm{D}\pi_Y \circ \mathrm{D}\Upsilon^t \circ \pi_{E^-,N} : N \to N, \tag{3.13}$$

generated by $\mathrm{D}v_Y \circ \pi_{E^-,N}$ along solution curves $g(x(t))$. Both $\mathrm{D}\pi_Y|_{E^-}$ and $\pi_{E^-,N}$ are uniformly bounded, so the exponential estimates of $\mathrm{D}\Upsilon^t_-$ carry over to $\hat{\Psi}^t$ up to a constant factor:

$$\forall\, t \geq 0: \left\|\hat{\Psi}^t\right\| \leq \tilde{C}_-\, e^{\rho_- t} \quad \text{with} \quad \tilde{C}_- = 2\, \|\pi_{E^-}\|\, C_-. \tag{3.14}$$

We have thus constructed flows Φ^t and $\hat{\Psi}^t$ on X and N, respectively, that are generated by

$$v_X \circ g \quad \text{and} \quad \hat{A}(x) = \mathrm{D}v_Y(g(x)) \cdot \pi_{E^-,N}(g(x))$$

with a solution curve $x(t)$ of the vector field $v_X \circ g$ inserted. These flows are of the form (3.12) and satisfy exponential estimates (1.10) inherited from the invariant bundle splitting.

The vector field $v_X \circ g$ already has sufficient smoothness[5], but $\hat{A}$ is not smooth enough since the projection $\pi_{E^-_\cdot N}$ is only continuous. We construct $A \in C^{k,\alpha}_{b,u}$ as a smoothed approximation of $\mathrm{D}_y v_Y \circ g$. This term is C^0-close to $\hat{A}$, since

$$\left\| \hat{A} - \mathrm{D}_y v_Y \circ g \right\|_0 = \left\| (\mathrm{D}_x v_Y \circ g) \cdot \mathrm{D}\pi_X \cdot \pi_{E^-_\cdot N} \right\|_0 \le \| \mathrm{D}_x v_Y \circ g \|_0 \left\| \pi_{E^-_\cdot N} \right\|_0 \tag{3.15}$$

because $\mathrm{D}\pi_Y \cdot \pi_{E^-_\cdot N} = \mathbb{1}_N$ and $\| \mathrm{D}_x v_Y \circ g \|$ is small. Lemma 3.17 will imply that the flow Ψ^t of this approximation has exponential growth estimates close to those of $\hat{\Psi}^t$. The following lemma will be used to obtain A from $\mathrm{D}_y v_Y \circ g$. We apply it with $l = k - 1$ to obtain $A \in C^k_b(X; \mathcal{L}(Y))$ such that $\left\| A - \mathrm{D}_y v_Y \circ g \right\|_{k-1} \le \varepsilon(\nu)$. This lemma is a (strongly) simplified version of Theorem 2.38; the notation of l, k is reversed to match the context here.

Lemma 3.12 (Uniform smoothing of a vector bundle section) *Let (X, g) be a Riemannian manifold of bounded geometry and V a Banach space. Let $f \in C^l_{b,u}(X; V)$ be a section of the trivial vector bundle $\pi : X \times V \to X$.*

Then for any $k > l$ and $\varepsilon > 0$ there exists a smoothed function $\tilde{f} \in C^k_{b,u}(X; V)$ such that $\left\| \tilde{f} - f \right\|_l \le \varepsilon$. (The bounds on higher than derivatives will generally depend on ε.)

Proof We apply convolution smoothing of Lemma 2.34 in each chart of a cover of X and glue these together.

Let $0 < \delta_1 < \delta_2 < \delta_3$ and let $\{B(x_i; \delta_2)\}_{i \ge 1}$ be a uniformly locally finite cover of X obtained from Lemma 2.16, such that the δ_1-sized sets already cover X, and the δ_3-sized sets still have normal coordinate charts. Lemma 2.17 yields a uniform partition of unity $\sum_{i \ge 1} \chi_i$ subordinate to this cover.

In each chart $B(x_i; \delta_2)$ we apply Lemma 2.34 to f with $r = \delta_2$ and $2\,\delta r \le \delta_3 - \delta_2$. We obtain $\tilde{f}_i \in C^k_{b,u}$ on each chart with uniformly bounded C^k-norms and $\left\| \tilde{f}_i - f \right\|_l$ can be made as small as required by choosing the parameter ν small. We glue these together to one function

$$\tilde{f} = \sum_{i \ge 1} \chi_i \, \tilde{f}_i$$

defined globally on X with the functions $\chi_i \in C^k_{b,u}$. Together with the uniform bound on the number of charts in the cover that intersect any one point, this guarantees that $\tilde{f}$ satisfies estimates equivalent to those of the $\tilde{f}_i$. Note that $\left\| \tilde{f} \right\|_l$ does not depend on the smoothing parameter ν, but the higher derivative norms do. □

---

[5] It may seem impossible to define a C^k vector field v on the tangent bundle $\mathrm{T}M$ of a C^k manifold M since $\mathrm{T}M \in C^{k-1}$. See [PT77, App. 1] or [PT83, p. 398] for a method to endow an invariant submanifold $M \in C^k$ with a compatible topology that makes $v|_M \in C^k$. We effectively used this in our definition of $v_X \circ g \in C^k$.

Remark 3.13 (*On loss of smoothness*). We must carefully construct the system 3.11 in order not to lose one degree of smoothness, while at the same time retaining exponential growth rates and proximity estimates.

The invariant complementary bundle E^- is only continuous, while the normal bundle of M is only C^{k-1}, even if disguised in coordinate expressions. We use the linearization at $y = 0$, but not directly, since $\mathrm{D}_y v_y(\,\cdot\,, 0) \in C^{k-1,\alpha}_{b,u}$ artificially decreases the smoothness as well. The loss of smoothness in [Sak90] occurs for these reasons. Note that even though we retain $C^{k,\alpha}$ smoothness by a convolution smoothing, this does not preserve higher than C^{k-1} bounds. This seems to be an artifact of the proof, inherent to the partial linearization along Y.

In the proof of Theorem 3.1 we construct a smoother, approximate manifold M_σ exactly to circumvent these problems. In the trivial bundle setting of Theorem 3.2 then, we must be careful not to pick a representation that reintroduces this loss of smoothness. On the other hand, we do not seem to obtain optimal results in the sense that we require $\|h\|_2$ small, while the classical results in the compact case only require $M = \mathrm{Graph}(h) \in C^1$. Similarly, $h \in C^k_b$ with $k \geq 3$ is assumed in [Sak90, p. 50], while hypothesis H2 in [BLZ99, p. 987] is imposed to bound 'twisting' of the invariant manifold. This requirement seems closely related to our condition on h, and is necessary for the same reason as in our Theorem 3.1: to construct a tubular neighborhood of uniform size. I do not know whether these stronger assumptions can be weakened or removed. ◊

3.5 Growth Estimates for the Perturbed System

We shall finally put all the ingredients together to obtain exponential growth estimates for perturbed flows contained in the tubular neighborhood $\|y\| \leq \eta$ of $X \times Y$. We write the perturbed vector field $\tilde{v}$ on $X \times Y$ as

$$\begin{aligned} \dot{x} &= \tilde{v}_X(x, y), \\ \dot{y} &= \tilde{v}_Y(x, y) = A(x)\, y + \tilde{f}(x, y), \\ \text{where} \quad \tilde{f}(x, y) &= f(x, y) + \big(\tilde{v}_Y(x, y) - v_Y(x, y)\big). \end{aligned} \tag{3.16}$$

Let us assume that the conditions of Theorem 3.2 hold true. First, if $\rho_M = 0$, then for any fixed r we can always slightly increase[6] to $\rho_M > 0$, such that the growth rates (1.10) and spectral gap condition $\rho_- < -r\,\rho_M$ still hold true; this way, we get rid of degenerate exponentials in integrals. We have some 'spectral space' $\Delta\rho = r\,\rho_M - \rho_- > 0$ that we use to define modified exponential growth numbers

---

6 We have $\rho_M \geq 0$ from (1.10). Note that we are interested in $-\rho_M$ for the stable side of the spectrum. In the rest of this chapter, all exponential rates will be negative.

$$\rho_X = -\rho_M - \frac{\Delta\rho}{4}, \qquad C_X = 2\,\tilde{C}_M,$$
$$\rho_Y = \rho_- + \frac{\Delta\rho}{4}, \qquad C_Y = \tilde{C}_-. \tag{3.17}$$

This allows us to get all perturbed flows within these slightly modified growth rates, while we reserve another $\Delta\rho/2$ spectral space for later use, such as proving (higher order) differentiability. Note that both ρ_Y, ρ_X are negative since we focus on the stable normal bundle.

We first fix some notation to be used throughout the proof:

- C_v denotes the global $C^{k,\alpha}$ bound on v and $\tilde{v}$.
- $\varepsilon(\nu)$ denotes the perturbation size in Lemma 3.12 depending on the smoothing convolution parameter ν from Lemma 2.34, while $C_v(\nu)$ denotes the $C^{k,\alpha}$ bound on A, $\tilde{f}$, which may grow due to smoothing when $\nu \to 0$. We also have $\|A\|_{k-1}, \|\tilde{f}\|_{k-1} \le C_v$.
- ζ denotes a small bound both on the derivative of $\tilde{f}$ and on perturbations of the horizontal vector field on X, that is, we impose bounds

$$\sup_{\substack{x\in X\\ \|y\|\le\eta}} \left\|\mathrm{D}\tilde{f}(x,y)\right\| \le \zeta \quad \text{and} \quad \sup_{\substack{x\in X\\ \|y\|\le\eta}} \|\tilde{v}_X(x,y) - v_X(x,h(x))\| \le \zeta,$$

and the size of ζ will be controlled by δ, σ_1, ν, and η.

Let us point out here that multiple parameters must be chosen small, some dependent on other small parameters. The following graph shows all dependencies; an arrow indicates that the choice of a parameter influences the choice of the object pointed to.

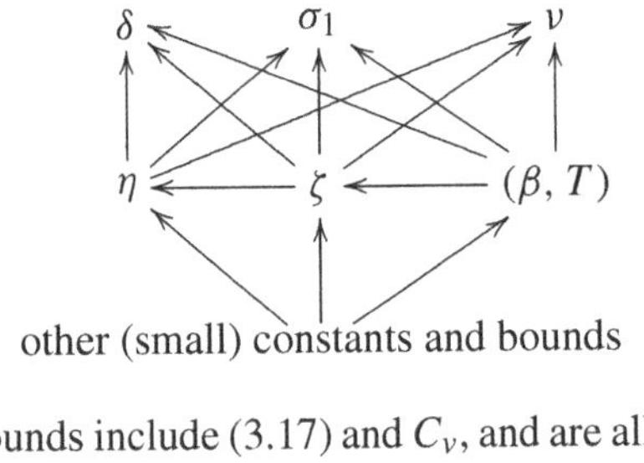

(3.18)

The constants and bounds include (3.17) and C_v, and are all fixed. Note that there are no circular dependencies, so we are free to choose any of these parameters smaller if necessary without the risk of having unsatisfiable constraints.

By invariance of $M = \{y = h(x)\}$ we have

$$v_Y(x,h(x)) = \frac{\mathrm{d}y}{\mathrm{d}t} = \mathrm{D}h(x)\cdot v_X(x,h(x)). \tag{3.19}$$

This can be used to estimate $\|v_Y(x,h(x))\| \le \sigma_1\, C_v$ and derived estimates, such as (taking the derivative with respect to x)

$$\begin{aligned}\|D_x v_Y \circ g\| &\le \|D_y v_Y\|\,\|Dh\| + \|D^2 h\|\,\|v_Y \circ g\| \\ &\quad + \|Dh\|\big(\|D_x v_X \circ g\| + \|D_y v_X \circ g\|\,\|Dh\|\big) \le 4\,C_v\,\sigma_1 \qquad (3.20)\end{aligned}$$

where $\sigma_1 \le 1$ has been assumed. Together with previous estimates, this leads to

$$\begin{aligned}\|f(x,0)\| &= \|v_Y(x,0) - A(x)\cdot 0\| \\ &\le \|Dh(x)\|\,\|v_X \circ g\|_0 \le \sigma_1\,C_v, \\ \|D_x f(x,y)\| &= \|D_x v_Y(x,y) - DA(x)\,y\| \\ &\le \|D_x v_Y(x,y) - D_x v_Y(x,h(x))\| + \|D_x v_Y \circ g\|_0 \\ &\quad + \Big(\|D(A - D_y v_Y \circ g)\|_0 + \|D_x D_y v_Y\|_0\Big)\,\|y\| \\ &\le \varepsilon_{D_x v_Y}(\eta + \sigma_1) + 4\,C_v\,\sigma_1 + \big(\varepsilon(\nu) + C_v\big)\,\eta, \\ \|D_y f(x,y)\| &= \|D_y v_Y(x,y) - A(x)\| \\ &\le \|D_y v_Y(x,y) - D_y v_Y(x,h(x))\| + \|D_y v_Y \circ g - A\|_0 \\ &\le \varepsilon_{D_y v_Y}(\eta + \sigma_1) + \varepsilon(\nu), \\ \|\tilde{v}_X(x,y) - v_X(x,h(x))\| &\le \|\tilde{v}_X(x,y) - v_X(x,y)\| + \|v_X(x,y) - v_X(x,h(x))\| \\ &\le \delta + \varepsilon_{v_X}(\eta + \sigma_1).\end{aligned}$$

Hence, $\|f(\cdot,0)\|_0$ can be made small independently of η, while $\|f\|_1$, $\|\tilde{v}_X(\cdot,y) - v_X \circ g\|_0 \le \zeta$ can be obtained for any $\zeta > 0$ depending on δ, η, σ_1, $\varepsilon(\nu)$, and the continuity moduli ε_{Dv_Y}, ε_{v_X}. So if we set

$$\zeta = 5\,C_v\,\sigma_1 + 2\,\varepsilon_{Dv_Y}(\eta + \sigma_1) + \big(\varepsilon(\nu) + C_v\big)\,\eta + \varepsilon(\nu) + \varepsilon_{v_X}(\eta + \sigma_1) + \delta, \quad (3.21)$$

then $\|f\|_1$, $\|\tilde{v}_X(\cdot,y) - v_X \circ g\|_0 \le \zeta$ hold and ζ is small when δ, σ_1, ν, η are.

We need the following result to control the C^{k-1} distance of the perturbed manifold $\tilde{M}$ to M.

Proposition 3.14 *For any $\varepsilon > 0$, the nonlinearity $\tilde{f}$ and its partial derivatives with respect to $x \in X$ can be bounded as*

$$\forall\, 0 \le i \le k-1 : \left\| D_x^i \tilde{f} \right\| \le \varepsilon \qquad (3.22)$$

by choosing η, ν, σ_1, $\|h\|_k$, and $\|\tilde{v} - v\|_{k-1}$ small enough.

The idea of the proof is the following. If M is described exactly by $h(x) \equiv 0$, then by invariance we have $v_Y(x,0) \equiv 0$ (cf. (3.19)), hence $D_x^i v_Y(x,0) \equiv 0$ as well. We adapt the proof to incorporate small perturbations introduced by the nonzero function h and the convolution smoothing of A.

Proof Note that $\tilde{f}$ is defined by (3.16) and (3.11) as

$$\tilde{f}(x,y) = v_Y(x,y) - A(x)\cdot y + \big[\tilde{v}_Y(x,y) - v_Y(x,y)\big], \qquad (3.23)$$

where A is defined as a convolution smoothing of $\mathrm{D}_y v_Y \circ g$ such that $\|A - \mathrm{D}_y v_Y \circ g\|_{k-1} \leq \varepsilon(\nu)$. The term in brackets obviously becomes small when $\|\tilde{v} - v\|_{k-1}$ does. For the second term note that $\|y\| \leq \eta$, while $\|A\|_{k-1}$ is bounded close to $\|\mathrm{D}_y v_Y \circ g\|_{k-1}$, which in turn can be estimated by $\|\mathrm{D}_y v_Y\|_{k-1} \leq C_v$ and $\|h\|_{k-1}$ after application of Proposition C.3.

For the first term in (3.23) we use the continuity modulus of $\mathrm{D}_x^i v_Y$ to estimate

$$\|\mathrm{D}_x^i v_Y(x, y)\| \leq \|\mathrm{D}_x^i v_Y(x, h(x))\| + \varepsilon_{\mathrm{D}_x^i v_Y}\big(\|y - h(x)\|\big),$$

while $\|y - h(x)\| \leq \eta + \sigma_1$. We insert (3.19) and apply Proposition C.3 another time to obtain

$$\mathrm{D}_x^i v_Y(x, h(x)) = \mathrm{D}_x^i\big[\mathrm{D}h(x) \cdot v_X(x, h(x))\big] - \sum_{\substack{l \geq 0, m \geq 1 \\ l+m \leq i}} \mathrm{D}_x^l \mathrm{D}_y^m v_Y(x, h(x)) \cdot P_{m,i-l}\big(\mathrm{D}^\bullet h(x)\big).$$

This expression can be made small since $\|\mathrm{D}_x^l \mathrm{D}_y^m v_Y\| \leq C_v$ and each term contains at least one factor $\mathrm{D}^j h(x)$ for some $0 \leq j \leq k$. □

Remark 3.15 Note that we cannot improve the result to a C^k size estimate, since $\|A\|_k \leq C(\nu)$ may grow with $\nu \to 0$, while compensating this by choosing η smaller would introduce a circular dependency in (3.18). ◊

As the next step, we will derive exponential growth estimates for the perturbed system (3.16). More generally, we consider the horizontal flow Φ_y and vertical, linear flow Ψ_x generated by

$$\dot{x} = \tilde{v}_X(x, y), \tag{3.24a}$$
$$\dot{y} = A(x)\, y, \tag{3.24b}$$

with specific curves $y : I \to Y$ and $x : I \to X$ substituted, respectively. The following series of lemmas and propositions show that these flows are small perturbations of the flows of (3.12) and satisfy exponential growth rates (3.17). We prove the nonlinear case on X and the linear case on Y separately, since we use C^1 smoothness for the nonlinear case, while only continuity can be assumed for the linear case.

Lemma 3.16 **(Growth estimates for a perturbed system)** *Let X be a Riemannian manifold and let the system $\dot{x} = v(t, x)$ with $v, \mathrm{D}_x v \in C_b^0$ have flow Φ with exponential growth estimate*

$$\forall\, x_0 \in X,\ t \leq t_0 : \|\mathrm{D}\Phi(t, t_0, x_0)\| \leq C\, e^{\rho(t-t_0)}. \tag{3.25}$$

Let $\tilde{v} = v + r$ be a perturbed system generating a flow $\tilde{\Phi}$. For each $\tilde{\rho} < \rho$ and $\tilde{C} > C$, there exists a $\delta > 0$, such that if $\|r\|_0$, $\|\mathrm{D}_x r\|_0 < \delta$, then $\tilde{\Phi}$ satisfies the growth estimate (3.25) with $\tilde{\rho}$ and $\tilde{C}$ inserted.

Note that this lemma is formulated in backward time.

Proof Choose $T > 0$ sufficiently large such that $\tilde{C}\, e^{\rho(-T)} \leq e^{\tilde{\rho}(-T)}$. By continuous dependence of the solutions of differential equations on parameters (see Theorem A.6 and Remark A.7), a C^1 small perturbation r results in a C^1 small perturbed flow $\tilde{\Phi}$ on compact time intervals $-T \leq t - t_0 \leq 0$. This result is uniform in t_0, t when $v, r \in C_b^1$, where differentiation is understood with respect to x only. Hence we obtain

$$\sup_{\substack{x_0 \in X \\ -T \leq t - t_0 \leq 0}} \|\mathrm{D}\tilde{\Phi}(t, t_0, x_0)\|\, e^{-\rho(t-t_0)} \leq \tilde{C}$$

if δ is chosen sufficiently small. Writing $t - t_0 = -(n\,T + \tau)$ with $n \in \mathbb{N},\ \tau \in [0, T)$, we use the group property of the flow to obtain

$$\|\mathrm{D}\tilde{\Phi}(t, t_0, x_0)\| \leq \left(\tilde{C}\, e^{\rho(-T)}\right)^n \tilde{C}\, e^{\rho(-\tau)} \leq e^{\tilde{\rho}\, n(-T)}\, \tilde{C}\, e^{\rho(-\tau)} \leq \tilde{C}\, e^{\tilde{\rho}(t-t_0)}. \qquad \square$$

Lemma 3.17 **(Perturbation of linear flow)** *Let Y be a Banach space and let $A \in C_b^0\big(\mathbb{R}; \mathcal{L}(Y)\big)$ generate a flow $\Psi(t, t_0)$ with growth estimate*

$$\forall\, t \geq t_0 : \|\Psi(t, t_0)\| \leq C\, e^{\rho(t-t_0)}. \tag{3.26}$$

Let $\tilde{\rho} > \rho$ be given and set $\delta = \frac{\tilde{\rho}-\rho}{C} > 0$. If $B \in C_b^0\big(\mathbb{R}; \mathcal{L}(Y)\big)$ is globally bounded by δ, then the flow $\tilde{\Psi}(t, t_0)$ of $\tilde{A}(t) = A(t) + B(t)$ satisfies (3.26) with $\tilde{\rho}$ inserted.

Proof The variation of constants integral equation for $\tilde{\Psi}$ is

$$\tilde{\Psi}(t, t_0) = \Psi(t, t_0) + \int_{t_0}^{t} \Psi(t, \tau)\, B(\tau)\, \tilde{\Psi}(\tau, t_0)\mathrm{d}\tau. \tag{3.27}$$

We shall prove the estimate for $\tilde{\Psi}$ with an approach inspired by Gronwall's lemma. Note that our variation of constants formula (3.26) is slightly different from the standard context of Gronwall's lemma, since we do not have a bound for A.

We denote by $\psi(t, t_0) = C\, e^{\rho(t-t_0)}$ the bound on Ψ. Now $\tilde{\psi}(t, t_0) = C\, e^{\tilde{\rho}(t-t_0)}$ satisfies the integral equation

$$\tilde{\psi}(t, t_0) = \psi(t, t_0) + \int_{t_0}^{t} \psi(t, \tau)\, \delta\, \tilde{\psi}(\tau, t_0)\mathrm{d}\tau \tag{3.28}$$

when $\delta\, C = \tilde{\rho} - \rho$. We verify this by calculating the right-hand side:

$$
\begin{aligned}
C\, e^{\rho(t-t_0)} + \int_{t_0}^{t} C\, e^{\rho(t-\tau)}\, \delta\, C\, e^{\tilde{\rho}(\tau-t_0)} \mathrm{d}\tau \\
&= C\, e^{\rho(t-t_0)} \Big[1 + \delta\, C \int_{t_0}^{t} e^{(\tilde{\rho}-\rho)(\tau-t_0)} \mathrm{d}\tau \Big] \\
&= C\, e^{\rho(t-t_0)} \Big[1 + \frac{\delta\, C}{\tilde{\rho}-\rho} \big(e^{(\tilde{\rho}-\rho)(t-t_0)} - 1 \big) \Big] \\
&= C\, e^{\rho(t-t_0)}\, e^{(\tilde{\rho}-\rho)(t-t_0)} \\
&= C\, e^{\tilde{\rho}(t-t_0)}.
\end{aligned}
$$

Next, we prove by contradiction that

$$
\|\tilde{\Psi}(t, t_0)\| \leq \tilde{\psi}(t, t_0).
$$

Thus, let

$$
t_1 = \inf\, \{t \in \mathbb{R} | t \geq t_0 \text{ and } \|\tilde{\Psi}(t, t_0)\| > \tilde{\psi}(t, t_0).
$$

Note that $\tilde{\Psi}$ is the solution of a differential equation, hence continuous. We write $\|\tilde{\Psi}(t, t_0)\| = \tilde{\psi}(t, t_0) + f(t)$, so we may assume that $f(t) \leq 0$ for $t \in [t_0, t_1]$, but there exist $t \in (t_1, t_2]$ arbitrary close to t_1 such that $f(t) > 0$. Let $f|_{[t_1,t_2]}$ attain its supremum at t, thus we have

$$
\sup_{[t_1,t]} f = f(t) > 0.
$$

We insert these estimates into the integral equality (3.27) and obtain

$$
\begin{aligned}
\|\tilde{\Psi}(t, t_0)\| = \tilde{\psi}(t, t_0) + f(t) &\leq \psi(t, t_0) + \int_{t_0}^{t} \psi(t, \tau)\, \delta \left(\tilde{\psi}(\tau, t_0) + f(\tau) \right) \mathrm{d}\tau \\
&\leq \tilde{\psi}(t, t_0) + \int_{t_1}^{t} \psi(t, \tau)\, \delta\, f(\tau) \mathrm{d}\tau \\
&\leq \tilde{\psi}(t, t_0) + (t_1 - t)\, \delta \sup_{\tau \in [t_1,t]} \psi(t, \tau) \sup_{[t_1,t]} f,
\end{aligned}
$$

where we used that $\tilde{\psi}(t, t_0)$ satisfies (3.28) and that $f|_{[t_0,t_1]} \leq 0$. Now we choose t_2 and therefore t sufficiently small that $(t - t_1)\, \delta\, \sup_{\tau \in [t_1,t]} \psi(t, \tau) \leq q < 1$, which leads to the contradiction

$$
f(t) \leq q \sup_{[t_1,t]} f < f(t). \qquad \square
$$

Proposition 3.18 **(Perturbation of X-flow estimate)** *If δ, η, σ_1 are sufficiently small, then the flow of (3.24a) satisfies the modified exponential growth estimates (3.17) for any $y \in \overline{B_\eta(\mathbb{R}; Y)}$ inserted.*

Here $\overline{B_\eta(\mathbb{R}; Y)}$ denotes the closed ball of radius η in the space of bounded continuous functions $\mathbb{R} \to Y$.

Proof Define the non-autonomous system $v(t, x) = \tilde{v}_X(t, x, y(t))$. This system is a C^1 small perturbation of $(v_X \circ g)(x)$, uniformly in t:

$$\begin{aligned}
\|v(t,x) - (v_X \circ g)(x)\| &\leq \|\tilde{v}_X(x, y(t)) - v_X(x, y(t))\| + \|v_X(x,(t)) - v_X(x, h(x))\| \\
&\leq \delta + \varepsilon_{\mathrm{D}v}(\eta + \sigma_1), \\
\|\mathrm{D}_x v(t,x) - \mathrm{D}(v_X \circ g)(x)\| &\leq \|\mathrm{D}_x \tilde{v}_X(x, y(t)) - \mathrm{D}_x v_X(x, y(t))\| \\
&\quad + \|\mathrm{D}_x v_X(x, y(t)) - \mathrm{D}_x v_X(x, h(x))\| \\
&\leq \delta + \varepsilon_{\mathrm{D}v}(\eta + \sigma_1),
\end{aligned}$$

where $\varepsilon_{\mathrm{D}v}$ denotes the uniform continuity modulus of v and its first derivative, which can be made small by choice of η, σ_1. We apply Lemma 3.16 to obtain exponential growth numbers C_X, ρ_X for (3.24a) by choosing $\delta + \varepsilon_{\mathrm{D}v}(\eta + \sigma_1)$ sufficiently small. □

The following definition and lemma for flows on X are again formulated in backward time, similar to Lemma 3.16.

Definition 3.19 **(Approximate solution)** Let X be a Riemannian manifold and $v(t, x)$ a time-dependent vector field on X. We call a continuous curve $x : \mathbb{R} \to X$ a (β, T)-approximate solution of v if for each interval $[t_2, t_1] \subset \mathbb{R}$ with $t_1 - t_2 \leq T$ and associated exact solution curve ξ of v with initial condition $\xi(t_1) = x(t_1)$, it holds that

$$\sup_{t_2 \leq t \leq t_1} d(x(t), \xi(t)) < \beta. \tag{3.29}$$

It would have been easier to define approximate solutions as C^1 curves x such that $\|\dot{x}(t) - v(t, x(t))\| < \beta$. We shall want to work with C^0-norms, though, and C^1 curves do not form a complete space under such norms. We use this continuous curve definition to avoid any complications associated with non-completeness. We still have the following result, as a discretized variant on variation by constants estimates.

Lemma 3.20 **(Growth of approximate solutions)** *Let X be a Riemannian manifold, $v(t, x)$ a time-dependent vector field on X, and x a (β, T)-approximate solution of v. Assume that v generates a flow Φ^{t,t_0} that satisfies the exponential growth estimate (3.25) with $C \geq 1$, $\rho < 0$. Let ξ_0 denote the exact solution of v with initial condition $\xi_0(0) = x(0)$.*

Then the distance $d_\rho(x, \xi_0)$ is finite on the interval $(-\infty, 0]$, and explicitly bounded by

$$d_\rho(x, \xi_0) \le \beta\Big(1 + \frac{C}{1 - e^{\rho T}}\Big). \tag{3.30}$$

Proof Let ξ_i with $i \in \mathbb{N}$ be the associated exact solutions of x that satisfy (3.29) on the interval $[-(i+1)T, -i\,T]$. We have $d\big(\xi_i(-(i+1)T), \xi_{i+1}(-(i+1)T)\big) < \beta$. Hence, $d_\rho(\xi_i, \xi_{i+1}) < \beta\, C\, e^{\rho(i+1)T}$ on the interval $(-\infty, (i+1)T]$ by the exponential growth estimate.

Thus on each interval $[-(i+1)T, -i\,T]$ we can use the triangle inequality to estimate

$$\begin{aligned} d_\rho(x, \xi_0) &\le d_\rho(x, \xi_i) + \sum_{j=0}^{i-1} d_\rho(\xi_j, \xi_{j+1}) \\ &< \beta\, e^{\rho\, i\, T} + \sum_{j=0}^{i-1} \beta\, C\, e^{\rho(j+1)T} \\ &\le \beta\Big(1 + \frac{C}{1 - e^{\rho T}}\Big). \end{aligned}$$

The union of all such intervals is $(-\infty, 0]$ hence (3.30) follows. □

Proposition 3.21 **(Perturbation of Y-flow estimate)** *Let x be a (β, T)-approximate solution to $v_X \circ g$. If T is sufficiently large and σ_1, ν, β are sufficiently small, then the flow Ψ_x of $A(x(t))$ has exponentially bounded growth as specified in (3.17), that is, $\|\Psi_x(t, t_0)\| \le C_Y\, e^{\rho_Y(t - t_0)}$ for all $t \ge t_0$.*

Proof Let $\tilde\rho = \frac{1}{2}(\rho_- + \rho_Y) < \rho_Y$ and choose $T > 0$ sufficiently large that $\tilde C_-\, e^{\tilde\rho\, T} \le e^{\rho_Y T}$. Let x_i be an exact solution to $v_X \circ g$ such that $\sup_{t \in [t_i, t_i + T]} d(x(t), x_i(t)) \le \beta$ per Definition 3.19, hence the flow $\hat\Psi$ of $\hat A(x_i(t))$ satisfies (3.14), that is, $\big\|\hat\Psi^t\big\| \le \tilde C_-\, e^{\rho_-\, t}$. We decompose $A(x(t)) = \hat A(x_i(t)) + B(t)$ and estimate

$$\begin{aligned} \|B(t)\| &= \|A(x(t)) - \hat A(x_i(t))\| \\ &\le \|A(x(t)) - A(x_i(t))\| + \|A(x_i(t)) - (\mathrm{D}_y v_Y \circ g)(x_i(t))\| \\ &\quad + \|(\mathrm{D}_y v_Y \circ g)(x_i(t)) - \hat A(x_i(t))\| \\ &\le \|\mathrm{D}A\|\, d(x(t), x_i(t)) + \varepsilon(\nu) + 4\, C_\nu\, \sigma_1\, 2\, \|\pi_{E^-}\|. \end{aligned}$$

Note that $\|A\|_1$ is bounded close to $\|\mathrm{D}_y v_Y\|_1 \le C_\nu$, and (3.15) and (3.20) were used to estimate the third term. We thus have $\|B(t)\| \le \delta$ for any $\delta > 0$ when σ_1, ν, β are sufficiently small. Hence by Lemma 3.17, we have $\|\Psi_x(\tau, \tau_0)\| \le \tilde C_-\, e^{\tilde\rho(\tau - \tau_0)}$ for any $\tau, \tau_0 \in [t_i, t_i + T]$.

Now we cover the interval $[t_0, t]$ by intervals $[t_0 + (i-1)T, t_0 + i\,T]$ with corresponding exact solutions x_i that approximate x. As in the proof of Lemma 3.16, we write $t - t_0 = n\,T + \tau$ and use the group property of the flow to obtain

$$\|\Psi_x(t, t_0)\| \leq \left(\tilde{C}_- e^{\tilde{\rho}\, T}\right)^n \tilde{C}_- e^{\tilde{\rho}\, \tau} \leq e^{\rho_Y\, n\, T}\, \tilde{C}_- e^{\rho_Y\, \tau} = \tilde{C}_- e^{\rho_Y (t - t_0)}.$$

We note that $C_Y = \tilde{C}_-$ to complete the proof. □

Using these results, we choose T sufficiently large and $\delta,\ \eta,\ \sigma_1,\ \beta,\ \nu$ sufficiently small that the modified flows $\mathrm{D}\Phi_y,\ \Psi_x$ satisfy exponential growth rates (3.17) when curves $y \in \overline{B_\eta(\mathbb{R}; Y)}$ and (β, T)-approximate solutions $x \in C^0(\mathbb{R}; X)$ are inserted.

Lemma 3.22 **(Variation of linear flow)** *Let X be a metric space and Y a Banach space and let $A \in C^\alpha_{b,u}(X; \mathcal{L}(Y))$ be a family of linear operators on Y that depends uniformly α-Hölder continuous on $x \in X$, with Hölder coefficient C_α and $0 < \alpha \leq 1$. Let Ψ_x denote the flow of A under a curve $x \in C(I; X)$, and assume that it satisfies the exponential growth condition (3.26).*

Then the variation of the flow satisfies the Hölder-like estimate

$$\|(\Psi_1 - \Psi_2)(t, \tau)\| \leq \frac{C_\alpha\, C^2}{-\alpha\, \rho}\, e^{\rho (t - \tau)}\, d_\rho(x_1, x_2)^\alpha\, e^{\alpha\, \rho\, \tau} \tag{3.31}$$

when $d_\rho(x_1, x_2)$ is finite.

Proof Let $d_\rho(x_1, x_2)$ be finite, let $\Psi_1,\ \Psi_2$ be the associated flows of A and denote $\Upsilon = \Psi_1 - \Psi_2$. We have for Υ the differential equation

$$\frac{\mathrm{d}}{\mathrm{d}t} \Upsilon(t, \tau) = A(x_1(t))\, \Upsilon(t, \tau) + \left[A(x_1(t)) - A(x_2(t))\right] \Psi_2(t, \tau), \qquad \Upsilon(t, t) = 0.$$

By variation of constants we obtain

$$\begin{aligned}
\|\Upsilon(t, \tau)\| &\leq \int_\tau^t \|\Psi_1(t, \sigma)\|\, C_\alpha\, d(x_1(\sigma), x_2(\sigma))^\alpha\, \|\Psi_2(\sigma, \tau)\| \mathrm{d}\sigma \\
&\leq C_\alpha\, C^2\, e^{\rho (t - \tau)} \int_\tau^t d_\rho(x_1, x_2)^\alpha\, e^{\alpha\, \rho\, \sigma} \mathrm{d}\sigma \\
&\leq \frac{C_\alpha\, C^2}{-\alpha\, \rho}\, e^{\rho (t - \tau)}\, d_\rho(x_1, x_2)^\alpha\, e^{\alpha\, \rho\, \tau}.
\end{aligned}$$ □

3.6 Existence and Lipschitz Regularity

We start with proving Lipschitz estimates for two mappings onto curves in X, Y, respectively. These mappings will be combined to a contraction mapping T. Its fixed points parametrized by $x_0 \in X$ will correspond to the unique solution curves of the modified system (3.16) that stay bounded.

Let $B^\rho(I;Y)$ denote the Banach space of exponentially bounded, continuous curves in Y on the interval $I = (-\infty, 0]$ and recall that in this chapter we always assume $\rho < 0$. Additionally, we denote by $B^\rho_\eta(I;Y) = B^\rho(I;Y) \cap B_\eta(I;Y)$ the subset of curves $y(t)$ which are moreover globally smaller than η. The closure of $B^\rho_\eta(I;Y)$ is given by $\bar{B}^\rho_\eta(I;Y) = B^\rho(I;Y) \cap \overline{B_\eta(I;Y)}$.

Proposition 3.23 *The space $\bar{B}^\rho_\eta(I;Y)$ is a closed subspace of the Banach space $B^\rho(I;Y)$, hence a complete metric space.*

Proof Consider the evaluation mapping $\mathrm{ev}_t : B^\rho(I;Y) \to Y : y \mapsto y(t)$. For each fixed $t \in I$ this is a continuous mapping as $\|y(t)\| \le \|y\|_\rho \, e^{\rho t}$ with $e^{\rho t}$ a finite number.

Let $R = \overline{B(0;\eta)}$ be the closed ball in Y, then we have

$$\bar{B}^\rho_\eta(I;Y) = \bigcap_{t \in I} \mathrm{ev}_t^{-1}(R)$$

as an intersection of closed preimages under ev_t, hence closed. □

For curves $x(t)$ in X, we cannot construct a similar space $B^\rho(I;X)$ as X is not a normed linear space. Instead we construct a (not necessarily complete) metric space. Let $\mathcal{B}^\rho(I;X) = \big(C^0(I;X), d_\rho\big)$ denote the space of continuous curves equipped with the metric (1.17) (which is allowed to take the value ∞) and let $\mathcal{B}^\rho_\beta(I;X)$ be the subset of curves $x \in C^0(I;X)$ that are (β, T)-approximate solutions to $v_X \circ g$ according to Definition 3.19. We suppress the dependence on T from the notation (note that T in the equation below is a completely different object); both β and T were fixed once and for all to fulfill the requirements of Proposition 3.21, we keep just the subscript β as a reminder and to distinguish from the space $\mathcal{B}^\rho(I;X)$. Let $\rho < \rho_X$. Then exact solutions of $v_X \circ g$ have finite d_ρ distance on I, and by Lemma 3.20 the distance of any two curves $x_1, x_2 \in \mathcal{B}^\rho_\beta(I;X)$ is finite, too.

We write $T = T_Y \circ \big(T_X\,,\ \mathrm{pr}_1\big)$ with

$$\begin{aligned} T &: \bar{B}^\rho_\eta(I;Y) \times X &&\to B^\rho_\eta(I;Y), \\ T_X &: \bar{B}^\rho_\eta(I;Y) \times X &&\to \mathcal{B}^\rho_\beta(I;X), \\ T_Y &: \mathcal{B}^\rho_\beta(I;X) \times B^\rho_\eta(I;Y) &&\to B^\rho_\eta(I;Y) \subset \bar{B}^\rho_\eta(I;Y), \end{aligned} \tag{3.32}$$

for any $\rho_Y < \rho < \rho_X$. The map T_X is defined by the flow Φ_y of $\tilde{v}_X(\,\cdot\,, y(t))$ with initial value $x_0 \in X$, that is,

$$T_X(y, x_0)(t) = \Phi_y(t, 0, x_0). \tag{3.33}$$

In [Hen81], the T_Y part of the contraction operator is indirectly defined by another contraction. Instead, here, we will set up T_Y as a direct mapping

$$T_Y(x, y)(t) = \int_{-\infty}^{t} \Psi_x(t, \tau)\, \tilde{f}(x(\tau), y(\tau))\mathrm{d}\tau, \tag{3.34}$$

where Ψ_x is the flow of $A(x(t))$. This should ease proving smoothness properties of T, which will subsequently imply smoothness of the invariant manifold.

Remark 3.24 Note that $\mathcal{B}^{\rho}_{\beta}(I; X)$ is not a Banach space or even a complete metric space. It will only appear as an intermediate space in the composition $T = T_Y \circ \big(T_X\,,\ \mathrm{pr}_1\big)$ though, so this does not affect the Banach fixed point arguments, as long as the mappings T_X, T_Y compose to a contraction T on $\bar{B}^{\rho}_{\eta}(I; Y)$ uniformly in the parameter $x_0 \in X$. ◊

The following two propositions show that the maps T_X, T_Y do indeed map into their specified codomains, when parameters are chosen sufficiently small.

Proposition 3.25 *If ζ is chosen such that (3.36) holds and δ, σ_1 are sufficiently small, then T_Y maps into $B_\eta(I; Y)$.*

Proof The conditions of Proposition 3.21 are satisfied for any $x \in \mathcal{B}^{\rho}_{\beta}(I; X)$, so the flow Ψ_x of system (3.24b) satisfies exponential growth estimates with numbers ρ_Y, C_Y.

Now, T_Y maps into $B_\eta(I; Y)$ since

$$\begin{aligned}\|T_Y(x, y)(t)\| &\le \int_{-\infty}^{t} \|\Psi(t, \tau)\| \left(\|\tilde{f}(x(\tau), 0)\| + \|\mathrm{D}_y \tilde{f}\|\, \|y(\tau)\|\right)\mathrm{d}\tau \\ &\le \frac{C_Y}{-\rho_Y}\,(C_v\,\sigma_1 + \delta + \zeta\,\eta)\end{aligned} \tag{3.35}$$

which can be made smaller than η by choosing ζ such that

$$\frac{C_Y\,\zeta}{-\rho_Y} \le \frac{1}{2} \tag{3.36}$$

holds, as well as δ, σ_1 sufficiently small. □

This shows that T_Y is well-defined in (3.32) after choosing ζ, δ, σ_1 possibly smaller. Note that the choice of ζ does not depend on any of the other small bounds. Similarly, we verify that T_X maps into $\mathcal{B}^{\rho}_{\beta}(I; X)$.

Proposition 3.26 *If δ, σ_1, and η are chosen sufficiently small, then T_X maps into $\mathcal{B}^{\rho}_{\beta}(I; X)$.*

Proof Let $y \in B^{\rho}_{\eta}(I; Y)$ and $x_0 \in X$. The curve $x = T_X(y, x_0)$ is generated by the vector field $\tilde{v}_X(\,\cdot\,, y(t))$ which is a small perturbation of $v_X \circ g$, since

$$\|\tilde{v}_X(\,\cdot\,, y(t)) - v_X \circ g\|_0 \le \zeta.$$

Let $t_2 - t_1 \le T$ and let Φ^t denote the flow of $v_X \circ g$. We apply the nonlinear variation of constants estimate (E.2) and obtain for $t \in [t_1, t_2]$

$$\begin{aligned} d\big(x(t), \Phi^{t-t_2}(x(t_2))\big) &\le \int_t^{t_2} \left\| \mathrm{D}\Phi^{t-\tau}(x(\tau)) \right\| \zeta \mathrm{d}\tau \\ &\le \int_t^{t_2} C_X \, e^{\rho_X (t-\tau)} \, \zeta \mathrm{d}\tau \\ &\le \frac{C_X \, \zeta}{-\rho_X} \, e^{\rho_X (t-t_2)}. \end{aligned}$$

Thus, if we choose ζ sufficiently small that

$$\frac{C_X \, \zeta}{-\rho_X} \, e^{\rho_X T} < \beta, \tag{3.37}$$

then x is (β, T)-approximated by the exact solution $\Phi^{t,t_2}(x(t_2))$ on the interval $[t_1, t_2]$ so $x \in \mathcal{B}^\rho_\beta(I; X)$. □

The basic argument for the Perron method is encoded in the following lemma: a *y-bounded* solution curve of the system (3.16) is equivalent to $y \in B_\eta(I; Y)$ being a fixed point of T, while this map T will be shown to be a contraction.

Lemma 3.27 (Cotton–Perron) *Let $x \in C(I; X)$, $y \in B_\eta(I; Y)$ bounded, and $x_0 \in X$. Then the following statements are equivalent:*

1. *the pair (x, y) is a solution curve for the modified system (3.16) with partial initial condition $x(0) = x_0$;*
2. *$y \in B_\eta(I; Y)$ is a fixed point of $T(\cdot\,, x_0)$ and $x = T_X(y, x_0)$.*

Proof The proof goes along the same lines as the classical Perron method for hyperbolic fixed points. As an intermediate step, we introduce the operator

$$\hat{T}_Y(x, y, t_0)(t) = \Psi_x(t, t_0)\, y(t_0) + \int_{t_0}^{t} \Psi_x(t, \tau)\, \tilde{f}(x(\tau), y(\tau)) \mathrm{d}\tau \tag{3.38}$$

and the following statement that is equivalent to those in the lemma:

3. *the pair (x, y) is a fixed point of $(T_X, \hat{T}_Y)$ for each $t_0 \in (-\infty, 0]$.*

Equivalence of *1* and *3* (with $t_0 = 0$) is a direct consequence of equivalence of differential and integral equations; the equation for y has been rewritten as a variation of constants integral with respect to the nonlinear term $\tilde{f}$. If (x, y) is a fixed point of $(T_X, \hat{T}_Y)$ for $t_0 = 0$, then this holds for any $t_0 \in (-\infty, 0]$. Note that the initial value for y is left unspecified in both statements.

We finish by proving the implications *3* ⇒ *2* ⇒ *1*. For the first we take the limit $t_0 \to -\infty$ in $\hat{T}_Y(x, y, t_0)$. Since Ψ_x decays exponentially and y is bounded, it follows that this limit is well-defined:

$$\forall\, t \in I : \lim_{t_0 \to -\infty} \hat{T}_Y(x, y, t_0)(t) = T_Y(x, y)(t).$$

Hence, a fixed point of $(T_X, \hat{T}_Y)$ is a fixed point of (T_X, T_Y). The last implication can readily be verified by calculating the time derivatives of $x = T_X(y, x_0)$ and $y = T_Y(x, y)$ to show that (x, y) is a solution of (3.16) with $x(0) = x_0$. □

Next, we prove that both T_X, T_Y are Lipschitz, while the Lipschitz constant of T_Y can be made arbitrarily small.

Lemma 3.28 *Let $\rho_Y < \rho \leq \rho_X$ and $0 < q_Y < 1$. If ζ is sufficiently small, then* $\mathrm{Lip}(T_Y) \leq q_Y$.

Proof Let (x_i, y_i), $i = 1, 2$ be curves from $\mathcal{B}^\rho_\beta(I; X) \times B^\rho_\eta(I; Y)$. Let Ψ_i be the corresponding flows of A along the curves x_i. Then the application of Lemma 3.22 in the Lipschitz case $\alpha = 1$ leads to a Lipschitz estimate on T_Y for any $\rho_Y < \rho \leq \rho_X$:

$$\begin{aligned}
&\|T_Y(x_1, y_1)(t) - T_Y(x_2, y_2)(t)\| \\
&\quad \leq \int_{-\infty}^{t} \|\Psi_1(t, \tau)\, \tilde{f}(x_1(\tau), y_1(\tau)) - \Psi_2(t, \tau)\, \tilde{f}(x_2(\tau), y_2(\tau))\| \mathrm{d}\tau \\
&\quad \leq \int_{-\infty}^{t} \|\Psi_1(t, \tau) - \Psi_2(t, \tau)\|\, \|\tilde{f}(x_1(\tau), y_1(\tau))\| \\
&\qquad + \|\Psi_2(t, \tau)\|\, \|\tilde{f}(x_1(\tau), y_1(\tau)) - \tilde{f}(x_2(\tau), y_2(\tau))\| \mathrm{d}\tau \\
&\quad \leq \int_{-\infty}^{t} \frac{C_v C_Y^2}{-\rho}\, e^{\rho_Y (t-\tau)}\, d_\rho(x_1, x_2)\, e^{\rho\,\tau}\, \|\tilde{f}\|_0 \\
&\qquad + C_Y\, e^{\rho_Y (t-\tau)} \|\mathrm{D}\tilde{f}\|_0 \left(d_\rho(x_1, x_2) + \|y_1 - y_2\|_\rho\right) e^{\rho\,\tau} \mathrm{d}\tau \\
&\quad \leq \zeta\, C \left(d_\rho(x_1, x_2) + \|y_1 - y_2\|_\rho\right) e^{\rho\, t}.
\end{aligned}$$

Here $C < \infty$ depends only on the constants and additional integration factors $(\rho - \rho_Y)^{-1}$ in the last integral, hence $\zeta\, C \leq q_Y$ when ζ is small enough. □

Lemma 3.29 *Let $\rho_Y < \rho < \rho_X$. If ζ, η are sufficiently small, then* $\mathrm{Lip}(T_X) \leq q_X$ *for some $q_X > 1$ independent of all small parameters.*

Proof Let $\xi_1, \xi_2 \in X$ and $y_1, y_2 \in B^\rho_\eta(I; Y)$. For $i = 1, 2$, define $v_i(t, \cdot) = \tilde{v}_X(\cdot, y_i(t))$ and let $x_i \in C^1(I; X)$ be a solution of the system v_i with initial condition ξ_i. We compare the systems v_1, v_2:

$$\|v_1(t, \cdot) - v_2(t, \cdot)\| \leq \left\|\mathrm{D}_y \tilde{v}_X\right\| \|y_1(t) - y_2(t)\| \leq C_v \|y_1(t) - y_2(t)\|.$$

The flow Φ_{y_1} has exponential growth numbers ρ_X, C_X. We view v_2 as a small perturbation of v_1 and apply the nonlinear variation of constants estimate (E.2) to obtain

$$\begin{aligned} d(x_1(t), x_2(t)) &\le C_X\, e^{\rho_X t}\, d(\xi_1, \xi_2) + \int_t^0 C_X\, e^{\rho_X (t-\tau)}\, C_v\, \|y_1(\tau) - y_2(\tau)\| \mathrm{d}\tau \\ &\le C_X\, e^{\rho_X t}\, d(\xi_1, \xi_2) + C_X\, C_v\, \|y_1 - y_2\|_\rho \int_t^0 e^{\rho_X (t-\tau)}\, e^{\rho\,\tau} \mathrm{d}\tau \\ &\le C_X\, e^{\rho_X t}\, d(x_1, x_2') + \frac{C_X\, C_v}{\rho_X - \rho} \|y_1 - y_2\|_\rho\, e^{\rho_X t}. \end{aligned}$$

Now

$$\begin{aligned} d_\rho\big(T_X(y_1, \xi_1), T_X(y_2, \xi_2)\big) &\le \sup_{t \le 0}\; C_X\, d(\xi_1, \xi_2)\, e^{(\rho_X - \rho)\, t} + \frac{C_X\, C_v}{\rho_X - \rho} \|y_1 - y_2\|_\rho\, e^{(\rho_X - \rho)\, t} \\ &\le C_X\, d(\xi_1, \xi_2) + \frac{C_X\, C_v}{\rho_X - \rho} \|y_1 - y_2\|_\rho \end{aligned}$$

exhibits a Lipschitz constant q_X for T_X that does not depend on any of the small parameters. □

Proposition 3.30 (Extension of solution is bounded in Y**)** *Let* $(x, y)(t)$ *be a solution of the perturbed system (3.16) satisfying* $y \in B_\eta(I; Y)$ *for* $t \le 0$*. For* ζ, δ, σ_1 *sufficiently small, the forward extension to* $t \ge 0$ *has* $y \in B_\eta(\mathbb{R}; Y)$*.*

Proof First of all, choose ζ, δ, σ_1 sufficiently small such that by (3.35), we have $y \in B_{\eta/2}(I; Y)$. Proceeding by contradiction, let t_0 be the first time after which $y(t)$ becomes larger than η, thus

$$t_0 = \sup\, \{t \in \mathbb{R} | \forall\, \tau \le t : \|y(\tau)\| \le \eta\}.$$

The curve $x(t)$ is a (β, T)-approximate solution to $v_X \circ g$ on the interval $(-\infty, t_0]$, so from Proposition 3.25 we conclude that $y \in B_{\eta/2}\big((-\infty, t_0]; Y\big)$. The continuity of y contradicts the assumption that t_0 is the supremum. □

Completing the Proof of Existence and Lipschitz Regularity

We finally put things together and prove that a unique persistent manifold $\tilde{M}$ exists and that it is Lipschitz.

Since T_X satisfies a fixed Lipschitz estimate, we can choose ζ small enough to obtain $q_X \cdot q_Y < 1$. Thus, T is a contraction on $\bar{B}_\eta^\rho(I; Y)$ for each fixed $\rho_Y < \rho < \rho_X$; ζ will depend on ρ though. According to Proposition 3.23, $\bar{B}_\eta^\rho(I; Y)$ is a complete metric space, so the Banach fixed point theorem shows that there is a unique $y \in \bar{B}_\eta^\rho(I; Y)$ fixed point of T; it holds moreover that $y \in B_\eta^\rho(I; Y)$. This contraction also depends (uniformly) on the parameter $x_0 \in X$, hence we obtain a fixed point map

$$\Theta^\infty : X \to B_\eta^\rho(I; Y), \tag{3.39}$$

satisfying the relation

$$\forall\, x_0 \in X : \Theta^\infty(x_0) = T(\Theta^\infty(x_0), x_0). \tag{3.40}$$

The superscript ∞ indicates that this map is obtained as a limit of applying the uniform contraction T. The parameter dependence in T is Lipschitz, so the map Θ^∞ will be Lipschitz as well.

By Proposition 3.30, the fixed point $y = \Theta^\infty(x_0)$ is bounded by η for all time and $B_\eta(I; Y) = B^\rho_\eta(I; Y)$ as sets, so y is the unique η-bounded solution with partial initial data $x(0) = x_0$. In combination with the evaluation map $y \mapsto y(0)$, we obtain the mapping

$$\tilde{h} : X \to B(0; \eta) \subset Y : x_0 \mapsto \Theta^\infty(x_0)(0). \tag{3.41}$$

Its graph $\tilde{M} = \text{Graph}(\tilde{h})$ is the unique invariant manifold of the modified system (3.16) and is Lipschitz as well.

Since both M and $\tilde{M}$ are described by graphs of small functions $X \to Y$, it follows that they are homeomorphic and $\|\tilde{h}\|_0 \leq \eta$. We can choose another, arbitrarily small η' instead. This requires us to choose smaller δ', σ_1' parameters as well. But as can be seen from (3.18), η does not depend on δ, σ_1, so the newly found $\|\tilde{h}'\|_0 \leq \eta'$ will actually be unique in the original η-sized neighborhood as well.

3.7 Smoothness

To study smoothness of Θ^∞, we can formally differentiate the fixed point relation (3.40) with respect to x_0 to obtain contractive mappings $T^{(k)}$ for the k-þorder derivatives of maps Θ and then apply the fiber contraction theorem, see Appendix D. If we assume that Θ satisfies (3.40), then Proposition C.3 shows that (at least formally)

$$\mathrm{D}^k\Theta(x_0) = \sum_{\substack{l,m\geq 0\\ l+m\leq k\\ (l,m)\neq(0,0)}} \mathrm{D}^l_y \mathrm{D}^m_{x_0} T(\Theta(x_0), x_0) \cdot P_{l,k-m}\big(\mathrm{D}^\bullet\Theta(x_0)\big), \tag{3.42}$$

which can be rewritten as a fiber contraction map on $\mathrm{D}^k\Theta$ by isolating that term $(l = 1,\ m = 0)$ on the right-hand side as

$$\mathrm{D}^k\Theta(x_0) = \mathrm{D}_y T(\Theta(x_0), x_0) \cdot \mathrm{D}^k\Theta(x_0) + \cdots$$

All the remaining terms are expressions in the lower order derivatives $\mathrm{D}^n\Theta(x_0)$ for $n < k$ only; these form the base space in the fiber contraction theorem.

The derivatives $\mathrm{D}^k\Theta$ and $\mathrm{D}^l_y\mathrm{D}^m_{x_0}T$ in (3.42) do not exist on the space $B^\rho_\eta(I; Y)$ as codomain, however. Indeed, if they did, we could have applied the implicit function theorem right away. Instead, the derivatives $\mathrm{D}^k\Theta$ are only well-defined on spaces[7] $B^{k\rho+\mu}(I;Y)$, where $\mu < 0$ is an arbitrarily small additional exponential growth rate. Derivatives of the maps T_X, T_Y do not exist at all. By using the fiber contraction theorem and interpreting the $\mathrm{D}^k T_X$, $\mathrm{D}^k T_Y$ as 'formal derivatives' in some appropriate way, we can still show, though, that the $\mathrm{D}^k\Theta$ are higher derivatives of Θ that converge to the derivatives of Θ^∞ under iteration of the fiber contraction maps (3.42). The gap condition $\rho_Y < r\,\rho_X$ will show up in the requirement that (3.42) is contractive for $k \le r$ (and finally $k + \alpha \le r$ when considering Hölder continuity). In case of uniform continuity (i.e. when $\alpha = 0$) we make use of the strict inequality to seize some of the spectral space left for the terms $\mu < 0$.

The interpretation of $\mathrm{D}T$ and its constituents $\mathrm{D}T_X$ and $\mathrm{D}T_Y$ as true derivatives is obstructed already by the fact that neither $B^\rho_\eta(I;Y)$ nor $\mathcal{B}^\rho_\beta(I;X)$ are smooth Banach manifolds[8], hence these can never be the (co)domain of differentiable maps. Thus, the chain rule

$$\mathrm{D}\Theta^{n+1} = \mathrm{D}_y T \cdot \mathrm{D}\Theta^n + \mathrm{D}_{x_0} T$$

cannot be used to conclude the existence of $\mathrm{D}\Theta^{n+1}$ from $\mathrm{D}\Theta^n$ by induction. On the other hand, we can find 'formal tangent bundles' of these spaces on which $\mathrm{D}T_X$, $\mathrm{D}T_Y$ are defined as 'formal derivatives', and we even have explicit formulas (3.43) for these maps. From here on we shall use the notation $\underset{\sim}{\mathrm{D}} f$ to indicate a formal derivative and $\mathrm{D}f$ to indicate that a function f is truly differentiable. We shall not make precise the notion of 'formal', but heuristically these formal objects can be seen as limits of well-defined real smooth manifolds and derivatives, see Sect. 3.7.7.

First, we outline the procedure of obtaining Θ^∞ as a truly differentiable map by careful manipulation of these formal derivatives. This is followed by the details of working out the definitions and estimates. Finally, we show how everything generalizes to higher derivatives. This last step adds more complexity, but requires no fundamentally new ideas.

Higher derivatives of functions involving variables or values in X need to be treated with some care, as these are not naturally defined. In such expressions, the derivatives are with respect to normal coordinates at the base point in domain and range, according to Definition 8.6. I should point the reader to Appendix C: it establishes the essential basic ingredient for this section on (higher) smoothness, namely how exponential growth estimates carry over to continuity and higher derivatives of the flow. Additionally, building on bounded geometry and Definition 2.9, a framework is set up to work with these notions on the manifold X.

---

7 Note that the spaces $B^{k\rho+\mu}(I;Y)$ are to be understood as the codomains of the maps Θ, hence the $\mathrm{D}^k\Theta$ as multilinear operators into these. The spaces Y_η play the same role in [Van89, Def. 3.10].

8 At least, they are not smooth Banach manifolds in a natural way, see the discussion in Sect. 3.7.4.

3.7.1 A Scheme to Obtain the First Derivative

The map T is not differentiable. Instead, we shall use the scheme below to obtain differentiability of Θ^∞. The sequence $\{\Theta^n\}_{n\geq 0}$ of maps $X \to B^\rho_\eta(I;Y)$ is defined by $\Theta^{n+1}(x_0) = T(\Theta^n(x_0), x_0)$ and $\Theta^0 \equiv 0$. We prove the differentiability of the Θ^n by induction and finally conclude that Θ^∞ is differentiable as well.

1. First, we propose candidate formal derivatives (3.43) of T_X, T_Y. These are obtained naturally by standard differentiation and variational techniques, postponing for the moment the question of which spaces these maps are well-defined on. We define $\underset{\sim}{\mathrm{D}}T$ in terms of the formal derivatives $\underset{\sim}{\mathrm{D}}T_X$ and $\underset{\sim}{\mathrm{D}}T_Y$.
2. The pair $(T, \underset{\sim}{\mathrm{D}}T)$ acts as a uniform fiber contraction on pairs of maps
$$(\Theta^n, \underset{\sim}{\mathrm{D}}\Theta^n) : \mathrm{T}X \to B^\rho_\eta(I;Y) \times B^{\rho+\mu}(I;Y)$$
when $\rho_Y < \rho + \mu \leq \rho < \rho_X$ holds, both in case of $\mu = 0$ and $\mu < 0$ small.
3. There are appropriate formal tangent bundles of the spaces $B^\rho_\eta(I;Y), \mathcal{B}^\rho_\beta(I;X)$ on which these formal derivatives are well-defined. Moreover, these formal tangent bundles can be endowed with a topology such that $\underset{\sim}{\mathrm{D}}T_X$, $\underset{\sim}{\mathrm{D}}T_Y$, and $\underset{\sim}{\mathrm{D}}T$ are uniformly continuous into bundles with slightly larger exponential growth rate $\rho+\mu$. Under appropriate assumptions (and with $\mu = \alpha\,\rho$) these formal derivatives are α-Hölder continuous.
4. The fiber contraction theorem D.1 can be applied. It follows from 2 that $\underset{\sim}{\mathrm{D}}T$ has a unique fixed point $\underset{\sim}{\mathrm{D}}\Theta^\infty : \mathrm{T}X \to \underset{\sim}{\mathrm{T}}B^\rho_\eta(I;Y)$, and from 3 that the map
$$\Theta \mapsto \underset{\sim}{\mathrm{D}}T(\Theta) \cdot \underset{\sim}{\mathrm{D}}\Theta^\infty : C^0\big(X; B^\rho_\eta(I;Y)\big) \to \Gamma_b\big(\mathcal{L}\big(\mathrm{T}X; \underset{\sim}{\mathrm{T}}B^{\rho+\mu}_\eta(I;Y)\big)\big)$$
into bounded sections of the bundle $\pi : \mathcal{L}\big(\mathrm{T}X; \underset{\sim}{\mathrm{T}}B^{\rho+\mu}_\eta(I;Y)\big) \to X$ is continuous. Thus we can conclude that $\underset{\sim}{\mathrm{D}}\Theta^n$ converges in $\Gamma_b\big(\mathcal{L}\big(\mathrm{T}X; B^{\rho+\mu}(I;Y)\big)\big)$ to the unique fixed point $\underset{\sim}{\mathrm{D}}\Theta^\infty$, simultaneously with $\Theta^n \to \Theta^\infty$. See (3.45) and (3.47) for precise definitions of these spaces. Moreover, $\underset{\sim}{\mathrm{D}}\Theta^\infty$ is uniformly or Hölder continuous.
5. There is a family of maps, given by restricting the domain I of curves,
$$T^{b,a} : B^\rho_\eta([a,0];Y) \times X \to B^\rho_\eta([b,0];Y)$$
that approximate T, and moreover these $T^{b,a}$ are differentiable maps between Banach manifolds whose derivatives $\mathrm{D}T^{b,a}$ approximate the formal derivative $\underset{\sim}{\mathrm{D}}T$.
6. With the continuous embedding $B^\rho(I;Y) \hookrightarrow B^{\rho+\mu}(I;Y)$ and the previous point, we show that if $\Theta^n : X \to B^{\rho+\mu}(I;Y)$ is differentiable, then

$$\underset{\sim}{\mathrm{D}}\Theta^{n+1} = \underset{\sim}{D}_y T \cdot \mathrm{D}\Theta^n + \underset{\sim}{\mathrm{D}}_{x_0} T$$

is the derivative of $\Theta^{n+1} : X \to B^{\rho+\mu}(I; Y)$.

7. Finally, we use Theorem D.2 to conclude that since the sequence Θ^n converges to Θ^∞ and its derivatives satisfy $\mathrm{D}\Theta^n \to \underset{\sim}{\mathrm{D}}\Theta^\infty$, it must hold that $\mathrm{D}\Theta^\infty = \underset{\sim}{\mathrm{D}}\Theta^\infty$ as a map into $B^{\rho+\mu}(I; Y)$.

In the subsequent sections we shall work out the details of this scheme. With some care, the same ideas generalize to higher derivatives.

3.7.2 Candidate Formal Derivatives

We first explicitly give the candidate mappings for the derivatives of T_X, T_Y. From now on, we will use shorthand notation $x_y(t) = T_X(y, x_0)(t) = \Phi_y(t, 0, x_0)$. The spaces that these maps act on will be made more precise in the following sections; δx, δy denote variations of curves $x \in \mathcal{B}^\rho_\beta(I; X)$ and $y \in B^\rho_\eta(I; Y)$, respectively, and $\delta x_0 \in \mathrm{T}_{x_0} X$.

$$\big(\underset{\sim}{\mathrm{D}}_{x_0} T_X(y, x_0)\,\delta x_0\big)(t) = \mathrm{D}\Phi_y(t, 0, x_0) \cdot \delta x_0 \tag{3.43a}$$

$$\big(\underset{\sim}{\mathrm{D}}_y T_X(y, x_0)\,\delta y\big)(t) = \int_t^0 \mathrm{D}\Phi_y(t, \tau, x_y(\tau))\,\mathrm{D}_y \tilde{v}_X(x_y(\tau), y(\tau))\,\delta y(\tau)\mathrm{d}\tau, \tag{3.43b}$$

$$\begin{aligned}\big(\underset{\sim}{\mathrm{D}}_x T_Y(x, y)\,\delta x\big)(t) = \int_{-\infty}^t &\Psi_x(t, \tau)\,\mathrm{D}_x \tilde{f}(x(\tau), y(\tau))\,\delta x(\tau)\\ &+ \big(\underset{\sim}{\mathrm{D}}_x \Psi_x \cdot \delta x\big)(t, \tau)\,\tilde{f}(x(\tau), y(\tau))\mathrm{d}\tau,\end{aligned} \tag{3.43c}$$

$$\big(\underset{\sim}{\mathrm{D}}_y T_Y(x, y)\,\delta y\big)(t) = \int_{-\infty}^t \Psi_x(t, \tau)\,\mathrm{D}_y \tilde{f}(x(\tau), y(\tau))\,\delta y(\tau)\mathrm{d}\tau, \tag{3.43d}$$

$$\big(\underset{\sim}{\mathrm{D}}_x \Psi_x \cdot \delta x\big)(t, \tau) = \int_\tau^t \Psi_x(t, \sigma)\,\mathrm{D}A(x(\sigma))\,\delta x(\sigma)\,\Psi_x(\sigma, \tau)\mathrm{d}\sigma. \tag{3.43e}$$

The correctness of these expressions pointwise in t can be checked by variation of constants and follows from Theorem E.2. Note also that the expressions above are linear in the variations δx, δy, δx_0. The map (3.43e) is only included in the list for its occurrence in (3.43c).

3.7.3 Uniformly Contractive Fiber Maps

We establish uniform boundedness of the formal derivative maps (**??**) as linear operators on δx, δy, and δx_0. The estimates are straightforward generalizations of those in Sect. 3.6. The operator norms are induced by $\|\cdot\|_\rho$ norms. We have the following list of estimates:

$$\|\underset{\sim}{\mathrm{D}}_{x_0} T_X(y, x_0)\| = \sup_{\substack{t\in I\\ \|\delta x_0\|=1}} \left\|\mathrm{D}\Phi_y(t, 0, x_0)\delta x_0\right\| e^{-\rho t} \le \sup_{t\in I} C_X\, e^{\rho_X t}\, e^{-\rho t} \le C_X,$$

$$\|\underset{\sim}{\mathrm{D}}_y T_X(y, x_0)\| \le \sup_{\substack{t\in I\\ \|\delta y\|_\rho=1}}$$

$$\int_t^0 \left\|\mathrm{D}\Phi_y(t, \tau, x_y(\tau))\, \mathrm{D}_y \tilde{v}_X(x_y(\tau), y(\tau))\, \delta y(\tau)\right\| \mathrm{d}\tau \cdot e^{-\rho t}$$

$$\le \sup_{t\in I} \int_t^0 C_X\, e^{\rho_X(t-\tau)}\, C_v\, e^{\rho(\tau-t)} \mathrm{d}\tau \le \frac{C_X\, C_v}{\rho_X - \rho},$$

$$\|\underset{\sim}{\mathrm{D}}_y T_Y(x, y)\| \le \sup_{\substack{t\in I\\ \|\delta y\|_\rho=1}} \int_{-\infty}^t \left\|\Psi_x(t, \tau)\, \mathrm{D}_y \tilde{f}(x(\tau), y(\tau))\, \delta y(\tau)\right\| \mathrm{d}\tau \cdot e^{-\rho t}$$

$$\le \sup_{t\in I} \int_{-\infty}^t C_Y\, e^{\rho_Y(t-\tau)}\, \zeta\, e^{\rho(\tau-t)} \mathrm{d}\tau \le \frac{C_Y\, \zeta}{\rho - \rho_Y},$$

$$\|(\underset{\sim}{\mathrm{D}}_x \Psi_x \cdot \delta x)(t, \tau)\| \le \int_\tau^t \|\Psi_x(t, \sigma)\, \mathrm{D}A(x(\sigma))\, \delta x(\sigma)\, \Psi_x(\sigma, \tau)\| \mathrm{d}\sigma$$

$$\le \int_\tau^t C_Y\, e^{\rho_Y(t-\sigma)}\, C_v\, \|\delta x\|_\rho\, e^{\rho\sigma}\, C_Y\, e^{\rho_Y(\sigma-\tau)} \mathrm{d}\sigma$$

$$\le \frac{C_Y^2\, C_v}{-\rho} \|\delta x\|_\rho\, e^{\rho_Y(t-\tau)}\, e^{\rho t},$$

$$\|\underset{\sim}{\mathrm{D}}_x T_Y(x, y)\| \le \sup_{\substack{t\in I\\ \|\delta x\|_\rho=1}} \int_{-\infty}^t \Big[\left\|\Psi_x(t, \tau)\, \mathrm{D}_x \tilde{f}(x(\tau), y(\tau))\, \delta x(\tau)\right\|$$

$$+ \|(\underset{\sim}{\mathrm{D}}_x \Psi_x \cdot \delta x)(t, \tau)\, \tilde{f}(x(\tau), y(\tau))\|\Big] \mathrm{d}\tau \cdot e^{-\rho t}$$

$$\le \sup_{t\in I} \int_{-\infty}^t C_Y\, e^{\rho_Y(t-\tau)}\, \zeta\, e^{\rho\tau} + \frac{C_Y^2\, C_v\, \zeta}{-\rho}\, e^{\rho_Y(t-\tau)}\, e^{\rho t} \mathrm{d}\tau \cdot e^{-\rho t}$$

$$\le \frac{C_Y\, \zeta}{\rho - \rho_Y} + \frac{C_Y^2\, C_v\, \zeta}{\rho \cdot \rho_Y}.$$

These estimates show that

$$\begin{aligned} \underset{\sim}{\mathrm{D}}_y T &= \underset{\sim}{\mathrm{D}}_x T_Y \cdot \underset{\sim}{\mathrm{D}}_y T_X + \underset{\sim}{\mathrm{D}}_y T_Y, \\ \underset{\sim}{\mathrm{D}}_{x_0} T &= \underset{\sim}{\mathrm{D}}_x T_Y \cdot \underset{\sim}{\mathrm{D}}_{x_0} T_X \end{aligned} \tag{3.44}$$

are bounded linear maps when $\rho_Y < \rho < \rho_X$. Since we have some spectral elbow room, we can first choose a value for ρ and then choose $\mu < 0$ sufficiently close to zero, such that this inequality holds both for ρ and $\rho + \mu$. If ζ is sufficiently small, then $\|\underset{\sim}{\mathrm{D}}_y T\| \le q < 1$ can be satisfied. This shows that $(T, \underset{\sim}{\mathrm{D}}T)$ is a uniform fiber contraction on $\delta y \in B^{\rho+\mu}(I; Y)$ over base curves $y \in B^{\rho}_{\eta}(I; Y)$, and with additional parameter $x_0 \in X$. It can also be viewed as a fiber mapping of maps $\underset{\sim}{\mathrm{D}}\Theta$ over base maps Θ. Let us define

$$\mathcal{S}_0 = C^0\big(X; B^{\rho}_{\eta}(I; Y)\big) \quad \text{and} \quad \mathcal{S}_1^{\mu} = \Gamma_b\big(\mathcal{L}(\mathrm{T}X; B^{\rho+\mu}(I; Y))\big), \tag{3.45}$$

where $\mathcal{S}_0$ is equipped with the supremum norm and $\mathcal{S}_1^{\mu}$ is interpreted as bounded sections of the bounded geometry bundle over X of linear maps between $\mathrm{T}X$ and the trivial bundle $\pi : X \times B^{\rho+\mu}(I; Y) \to X$, equipped with the (supremum/operator) norm

$$\|\underset{\sim}{\mathrm{D}}\Theta\| = \sup_{x_0 \in X} \|\underset{\sim}{\mathrm{D}}\Theta(x_0)\|_{\mathcal{L}(\mathrm{T}_{x_0} X; B^{\rho+\mu}(I; Y))}.$$

Then $(T, \underset{\sim}{\mathrm{D}}T)$ can also be viewed as a fiber mapping

$$\begin{aligned} (T, \underset{\sim}{\mathrm{D}}T) : \mathcal{S}_0 \times \mathcal{S}_1^{\mu} &\to \mathcal{S}_0 \times \mathcal{S}_1^{\mu}, \\ (\Theta, \underset{\sim}{\mathrm{D}}\Theta) &\mapsto \Big((x_0, \delta x_0) \mapsto \big(T(\Theta(x_0), x_0), \\ &\qquad \big[\underset{\sim}{\mathrm{D}}_y T(\Theta(x_0), x_0) \cdot \underset{\sim}{\mathrm{D}}\Theta(x_0) + \underset{\sim}{\mathrm{D}}_{x_0} T(\Theta(x_0), x_0)\big] \cdot \delta x_0\big)\Big). \end{aligned} \tag{3.46}$$

As such, it is again a uniform fiber contraction since the contraction was uniform in y and x_0 to begin with, and the supremum norm does not affect the contraction factor $q < 1$.

3.7.4 *Formal Tangent Bundles*

Derivatives of the maps T_X, T_Y should be defined between tangent bundles of the spaces $B^{\rho}_{\eta}(I; Y)$ and $\mathcal{B}^{\rho}_{\beta}(I; X)$. This is problematic for both spaces: $B^{\rho}_{\eta}(I; Y)$ is a subspace of the Banach space $B^{\rho}(I; Y)$, but it has empty interior. The restriction to (β, T)-approximate solutions of $v_X \circ g$ creates a similar problem for $\mathcal{B}^{\rho}_{\beta}(I; X)$, but here, construction of the tangent bundle faces an additional obstruction. There is no clear way to define local coordinates around a solution curve x. The obvious method would be by constructing a tubular neighborhood of x and represent nearby curves $\tilde{x}$ in the tubular neighborhood. But the metric on $\mathcal{B}^\rho_\beta(I; X)$ allows $\tilde{x}$ to diverge exponentially from x even if $d_\rho(x, \tilde{x})$ is small. Thus the tubular neighborhood would need to be of infinite size to contain $\tilde{x}(t)$ for all $t \in I$, which is generally not possible. Since any finite-size tubular neighborhood does not contain a full neighborhood of the curve x, we cannot use local coordinates to define tangent spaces.

Instead, we shall construct formal tangent bundles. These are just convenient spaces to model variations of curves on; they are natural extensions of true Banach tangent spaces, see Sect. 3.7.7. The primary role of these bundles is to introduce a topology that allows us to show that $\underset{\sim}{D}T$ is uniformly or Hölder continuous.

A formal tangent bundle of $B^\rho_\eta(I; Y)$ can be constructed rather easily: $B^\rho(I; Y)$ is a Banach space, so its tangent bundle is canonically identified as $\mathrm{T}B^\rho(I; Y) = B^\rho(I; Y) \times B^\rho(I; Y)$. We then define the formal tangent bundle of $B^\rho_\eta(I; Y)$ by restricting the base:

$$\underset{\sim}{\mathrm{T}}B^\rho_\eta(I; Y) = \mathrm{T}B^\rho(I; Y)|_{B^\rho_\eta(I;Y)} \cong B^\rho_\eta(I; Y) \times B^\rho(I; Y) \tag{3.47}$$

with induced topology and norm.

To define a formal tangent bundle of $\mathcal{B}^\rho_\beta(I; X)$, we consider variations δx of a curve x as sections of a pullback bundle: $\delta x \in \Gamma(x^*(\mathrm{T}X))$. That is, $\delta x \in C(I; \mathrm{T}X)$ is such that $\delta x(t) \in \mathrm{T}_{x(t)}X$ for each $t \in I$. We equip this space with the norm that is natural for our problem, namely

$$\|\delta x\|_\rho = \sup_{t\in I} \|\delta x(t)\| \, e^{-\rho t},$$

and denote it by $B^\rho(I; x^*(\mathrm{T}X)) = \big(\Gamma(x^*(\mathrm{T}X)), \|\cdot\|_\rho\big)$. The curves $\delta x \in B^\rho(I; x^*(\mathrm{T}X))$ form the formal tangent space over one curve $x \in \mathcal{B}^\rho_\beta(I; X)$. The complete formal tangent bundle is then defined as the coproduct over all curves $x \in \mathcal{B}^\rho_\beta(I; X)$,

$$\underset{\sim}{\mathrm{T}}\mathcal{B}^\rho_\beta(I; X) = \coprod_{x\in\mathcal{B}^\rho_\beta(I;X)} B^\rho(I; x^*(\mathrm{T}X)). \tag{3.48}$$

A curve δx lives above a specific base curve x, so there is no direct way of comparing two curves δx_1, δx_2 with different base curves x_1, x_2; (3.48) was constructed as a coproduct without topological structure. We add a topology based on parallel transport. This requires the base curves $x \in \mathcal{B}^\rho_\beta(I; X)$ to be differentiable, so we consider the bundle

$$\underset{\sim}{\mathrm{T}}\mathcal{B}^\rho_\beta(I; X)\big|_{C^1} \tag{3.49}$$

restricted to differentiable[9] base curves $x \in \mathcal{B}^\rho_\beta(I;X) \cap C^1$. Variational curves $\delta x \in B^\rho(I; x^*(\mathrm{T}X))$ are isometrically mapped onto curves $\widetilde{\delta x} \in B^\rho(I; \mathrm{T}_{x(0)}X) = \big(C^0(I; \mathrm{T}_{x(0)}X), \|\cdot\|_\rho\big)$ by

$$\tilde{\Pi}_x : \delta x \mapsto \widetilde{\delta x}, \qquad \widetilde{\delta x}(t) = \Pi(x|_0^t)^{-1}\,\delta x(t). \tag{3.50}$$

Let the normal coordinate radius δ_X be X-small as in Definition 2.8. If we now restrict all base curves x under consideration to a small neighborhood

$$U_{\bar{x}} = \{x \in \mathcal{B}^\rho_\beta(I;X) \cap C^1 | x(0) \in B(\bar{x};\delta_X), \tag{3.51}$$

i.e., the curves x that start in the open ball $B(\bar{x};\delta_X) \subset X$, then there exists a unique shortest geodesic $\gamma_{\bar{x},x(0)}$ from $x(0)$ to $\bar{x}$ for each $x \in U_{\bar{x}}$. Parallel transport along these geodesics induces a local trivialization[10] of $\mathrm{T}B(\bar{x};\delta_X)$. This in turn induces a local trivialization of $\underset{\sim}{\mathrm{T}}\mathcal{B}^\rho_\beta(I;X)\big|_{C^1}$:

$$\begin{array}{ccc} \underset{\sim}{\mathrm{T}}\mathcal{B}^\rho_\beta(I;X)|_{U_{\bar{x}}} & \xrightarrow{\quad\tau_{\bar{x}}\quad} & U_{\bar{x}} \times B^\rho(I;\mathrm{T}_{\bar{x}}X) \\ \Big\downarrow \pi & \swarrow p_1 & \\ \mathcal{B}^\rho_\beta(I;X)|_{C^1} \supset U_{\bar{x}} & & \end{array} \tag{3.52}$$

The trivialization map is given by $\tau_{\bar{x}}(x,\delta x) = \big(x, \Pi(\gamma_{\bar{x},x(0)}) \circ \tilde{\Pi}_x(\delta x)\big)$. The transition maps between overlapping local trivializations $U_{\bar{x}_1} \cap U_{\bar{x}_2} \neq \emptyset$ are induced by transition functions

$$\varphi_{2,1} : \big(B(\bar{x}_2;\delta_X) \cap B(\bar{x}_1;\delta_X)\big) \times \mathrm{T}_{\bar{x}_2}X \to \mathrm{T}_{\bar{x}_1}X : (\xi, v) \mapsto \Pi(\gamma_{\bar{x}_2,\xi} \circ \gamma_{\xi,\bar{x}_1}) \cdot v$$

between local trivializations of $\mathrm{T}B(\bar{x}_i;\delta_X)$. The map $\varphi_{2,1}$ is uniformly Lipschitz by Lemma 2.6 and linear in the fiber. This induces a Lipschitz continuous transition function $\tau_{\bar{x}_2} \circ \tau_{\bar{x}_1}^{-1}$ that depends on the base curve $x \in U_{\bar{x}_1} \cap U_{\bar{x}_2}$ only through $x(0) \in X$; this dependence is uniform since X has bounded geometry. Thus the bundle satisfies Definition 2.14, and the order of bounded geometry is actually equal to $k-2$ when X has k-th order bounded geometry.

---

9 This does not cause problems since T_X actually maps into curves $x \in C^1$. The fiber contraction theorem only requires that the base space has a globally attractive fixed point. Since the fixed point is a C^1 curve, we can simply restrict to this subset of curves.

10 We make the specific choice to trivialize $\mathrm{T}B(\bar{x};\delta_X)$ by parallel transport along geodesics. Any other trivialization with uniformly bounded transition maps would also suffice for our purposes and induce a trivialization of $\underset{\sim}{\mathrm{T}}\mathcal{B}^\rho_\beta(I;X)|_{U_{\bar{x}}}$ (see also the alternative viewpoint on this trivialization below). This explicit choice is somewhat natural in this context, though, and it shows that a trivialization with these properties does exist.

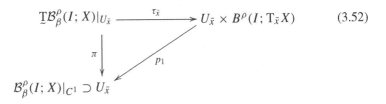

We endow the bundle $\underset{\sim}{\mathrm{T}}\mathcal{B}^{\rho}_{\beta}(I;X)\big|_{C^1}$ with the topology induced by these local trivializations. Note that this topology is induced by a locally defined distance function, so we can express uniform and Hölder continuity of maps on $\underset{\sim}{\mathrm{T}}\mathcal{B}^{\rho}_{\beta}(I;X)\big|_{C^1}$. That is, if $(x_1,\delta x_1)$ and $(x_2,\delta x_2)$ are elements of $\underset{\sim}{\mathrm{T}}\mathcal{B}^{\rho}_{\beta}(I;X)\big|_{C^1}$ such that $d_\rho(x_1,x_2)<\delta_X$, then the topology is induced by the locally defined distance function

$$d\big((x_1,\delta x_1),(x_2,\delta x_2)\big)=d_\rho(x_2,x_1)+\|\Pi(\gamma_{x_1(0),x_2(0)})\,\tilde{\Pi}_{x_2}(\delta x_2)-\tilde{\Pi}_{x_1}(\delta x_1)\|_\rho. \tag{3.53}$$

The transition functions $\tau_{\bar{x}_2}\circ\tau_{\bar{x}_1}^{-1}$ are uniformly Lipschitz, so they preserve uniform and Hölder continuity moduli up to a constant. Therefore, overlapping trivializations define the same topology on their intersection, with compatible local distances. To summarize, we have

Proposition 3.31 *The spaces $\underset{\sim}{\mathrm{T}}B^{\rho}_{\eta}(I;Y)$ and $\underset{\sim}{\mathrm{T}}\mathcal{B}^{\rho}_{\beta}(I;X)\big|_{C^1}$ are well-defined normed vector bundles of bounded geometry, and they have a (local) distance structure.*

The topologies introduced above allow us to express uniform and Hölder continuity of the maps (**??**). The topology on $\underset{\sim}{\mathrm{T}}B^{\rho}_{\eta}(I;Y)$ is clear and explicit from the topology on $B^{\rho}(I;Y)$. For $\underset{\sim}{\mathrm{T}}\mathcal{B}^{\rho}_{\beta}(I;X)\big|_{C^1}$ let $x\in\mathcal{B}^{\rho}_{\beta}(I;X)\cap C^1$ be a curve and $\delta x\in B^{\rho}(I;x^*(\mathrm{T}X))$ a variational curve at x. The topology is induced by the isometric representation

$$\widetilde{\delta x}=\Pi(\gamma_{\bar{x},x(0)})\cdot\tilde{\Pi}_x\cdot\delta x\in B^{\rho}(I;\mathrm{T}_{\bar{x}}X)$$

of δx. Uniform continuity of maps (**??**) that have $\underset{\sim}{\mathrm{T}}\mathcal{B}^{\rho}_{\beta}(I;X)\big|_{C^1}$ as (co)domain can thus be checked by switching to a local trivialization, that is, substitute

$$\delta x(t)=\Pi(x|_0^t)\cdot\Pi(\gamma_{x(0),\bar{x}})\cdot\widetilde{\delta x}(t)$$

and then use the known topology on $U_{\bar{x}}\times B^{\rho}(I;\mathrm{T}_{\bar{x}}X)$. In explicit calculations of continuity with respect to the base $\mathcal{B}^{\rho}_{\beta}(I;X)\cap C^1$, we shall thus add parallel transport terms such as those above to the maps (**??**) and let these act on $\widetilde{\delta x}\in B^{\rho}(I;\mathrm{T}_{\bar{x}}X)$.

Alternative Viewpoints

Instead of the immediate trivialization (3.52) of the bundle $\underset{\sim}{\mathrm{T}}\mathcal{B}^{\rho}_{\beta}(I;X)\big|_{C^1}$, we can also introduce an intermediate viewpoint that corresponds to only applying the parallel transport term $\tilde{\Pi}_x$, but not $\Pi(\gamma_{\bar{x},x(0)})$ in the local neighborhood $B(\bar{x};\delta_X)$. We view $\mathrm{ev}_0:\mathcal{B}^{\rho}_{\beta}(I;X)\to X$ as a bundle; this identifies $\underset{\sim}{\mathrm{T}}\mathcal{B}^{\rho}_{\beta}(I;X)\big|_{C^1}$ as a bundle over X as well, via $\mathrm{ev}_0\circ\pi$. Let $B^{\rho}(I;\mathrm{T}X)_X$ denote the space of (continuous, exponential growth) functions $\widetilde{\delta x}:I\to\mathrm{T}X$ such that $\pi\circ\widetilde{\delta x}$ is constant into X, viewed as a bundle over X.

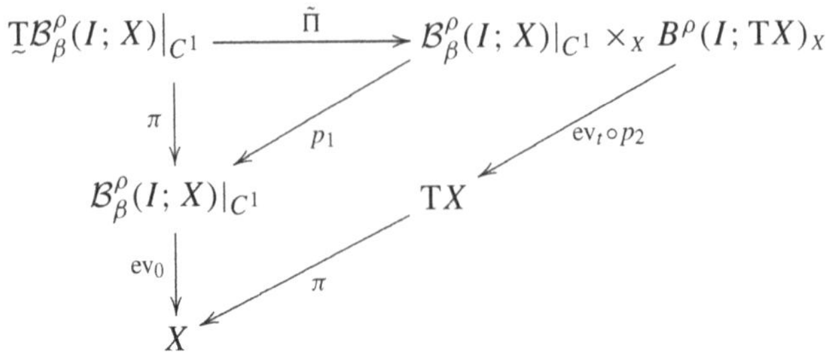

(3.54)

This commutative diagram shows that $\underset{\sim}{\mathrm{T}}\mathcal{B}^\rho_\beta(I;X)\big|_{C^1}$ can be identified via $\tilde{\Pi}$ with the fiber product bundle $\mathcal{B}^\rho_\beta(I;X)|_{C^1} \times_X B^\rho(I;\mathrm{T}X)_X$ over X. This identification is natural in the sense that no local trivialization of X or $\mathrm{T}X$ is used. The second component $B^\rho(I;\mathrm{T}X)_X$ of this bundle contains the variational curves $\widetilde{\delta x}$. This is a (nontrivial) bundle over X, but its projection onto the base $\pi \circ \mathrm{ev}_t : B^\rho(I;\mathrm{T}X)_X \to X$ factors through $\mathrm{T}X$. The fact that $\pi \circ \mathrm{ev}_t$ is constant for $t \in I$ simply expresses that each $\widetilde{\delta x} \in B^\rho(I;\mathrm{T}X)_X$ maps into a fixed tangent space $\mathrm{T}_\xi X$. This shows that a local trivialization $\sigma : \mathrm{T}X|_{B(\bar{x};\delta_X)} \to B(\bar{x};\delta_X) \times \mathbb{R}^n$ naturally lifts to a local trivialization

$$\tilde{\sigma} : B^\rho(I;\mathrm{T}X)|_{B(\bar{x};\delta_X)} \to B(\bar{x};\delta_X) \times B^\rho(I;\mathbb{R}^n).$$

We have chosen local trivializations of $\mathrm{T}X$ by parallel transport along geodesics, i.e. $\Pi(\gamma_{\bar{x},\xi})$, since this construction is compatible with the bounded geometry of X in the sense that trivialization chart transitions are C^k_b maps by Proposition 2.13.

We also introduce a reformulation of the topology on $\underset{\sim}{\mathrm{T}}\mathcal{B}^\rho_\beta(I;X)\big|_{C^1}$ using frames, as an alternative to the explicit formulation in terms of parallel transport above. This allows us to abstract away these ideas into a lighter notation in the next section and only recall the full details when required.

Let $e_{\bar{x}} : \mathbb{R}^n \to \mathrm{T}_{\bar{x}}X$ be a choice[11] of orthonormal frame at $\bar{x} \in X$. We can extend this to an orthonormal frame e on $\mathrm{T}B(\bar{x};\delta_X)$ by parallel transport of the frame $e_{\bar{x}}$ along geodesics emanating from $\bar{x}$. As a second step, we further extend the frame e along any curve $x \in U_{\bar{x}}$, again by parallel transport.[12]

We adopt the notation $v_f = f^{-1} \cdot v$ to express a vector $v \in \mathrm{T}_x X$ with respect to a frame f at x, and use this notation more generally on the tensor bundle of X. Now let v be a vector field and ω a one-form on X, then the construction of e above leads to

[11] The precise choice does not matter and will drop out in the final, relevant equations. The relative choice of frame along curves is what matters.

[12] Note that e does not define a (global) frame on $\mathrm{T}X$. The choice of frame at $x(t)$ depends not just on the point $x(t) \in X$, but on the whole curve $x \in \mathcal{B}^\rho_\beta(I;X)|_{C^1}$. Another curve $\tilde{x}$ with $x(t) = \tilde{x}(t)$ will generally induce a different frame in $\mathrm{T}_{x(t)}X$.

$$\begin{aligned} v(x(t))_e &= e_{\bar{x}}^{-1} \cdot \Pi(\gamma_{\bar{x},x(0)}) \cdot \Pi(x|_t^0) \cdot v(x(t)), \\ \omega(x(t))_e &= \omega(x(t)) \cdot \Pi(x|_0^t) \cdot \Pi(\gamma_{x(0),\bar{x}}) \cdot e_{\bar{x}}, \end{aligned} \tag{3.55}$$

and naturally extends to the tensor bundle of X.

3.7.5 *Continuity of the Fiber Maps*

We prove the uniform and Hölder continuous dependence on x, y, and x_0 of the maps (??) using a combination of techniques. One is the variation of constants formula to get expressions for the variation of flows when changing a parameter. Such variations require us to compare the variational curves over different base curves; for this, we use the topologies of the formal tangent bundles in Sect. 3.7.4, while we measure the variation of vector fields with the formulation of continuity via parallel transport in Proposition 2.13. Together these lead to holonomy terms along the base paths (see Fig. 3.7), in addition to the variation of constants terms that would simply occur in $\mathbb{R}^n$. These holonomy terms can be estimated with Lemma 2.19 and do not essentially alter the estimates.

We use Nemytskii operator techniques as laid out in Appendix B to conclude that functions such as A and f can be interpreted as uniformly continuous maps onto curves with some $\mu < 0$ exponential growth norm. Instead of uniform continuity, we can also obtain Hölder continuity if the original maps are Hölder continuous and if we view the Nemytskii operator as a mapping into a space with norm $\|\cdot\|_{\alpha\rho}$. In other words, we replace the uniform continuity modulus by the explicit α-Hölder continuity modulus. Hölder continuity precisely fits the problem, so in that case there is no need anymore to add a small $\mu < 0$ to the exponential growth norms.

One Example in Full Detail

As an example, let us consider continuity of the map (3.43c) with respect to $x \in \mathcal{B}_\beta^\rho(I; X)$, that is, $x \mapsto \underset{\sim}{D}_x T_Y(x, y)$. To be able to explicitly use the topology on $\underset{\sim}{T}\mathcal{B}_\beta^\rho(I; X)\big|_{C^1}$, we switch to a local trivialization neighborhood $U_{\bar{x}} \ni x_1, x_2$ as

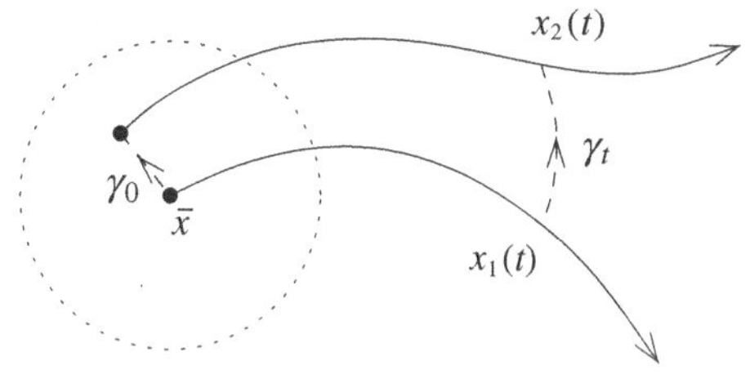

Fig. 3.7 Paths involved in the holonomy term

in (3.51). We choose $\bar{x} = x_1(0)$ to simplify expressions; any other choice for $\bar{x}$ can be obtained by a transition of trivialization charts. Let $\widetilde{\delta x} \in B^\rho(I; \mathrm{T}_{\bar{x}}X)$ be the representation of an arbitrary variational curve in the fiber of this trivialization.

Note that (3.43c) is defined in terms of (3.43e). We estimate continuity of the separate components and build towards the full expression. Let us first focus on the continuity of $x \mapsto \Psi_x(t, \tau)$, which is a map $\mathcal{B}^\rho_\beta(I; X) \to \mathcal{L}(Y)$ for fixed $t, \tau \in I$.

Proposition 3.32 *For any $\mu < 0$, the variation*

$$\Upsilon^{t,\tau} = \Psi_{x_2}(t, \tau) - \Psi_{x_1}(t, \tau) \tag{3.56}$$

of the linear flow Ψ_x on Y satisfies continuity estimate (3.57).

Proof We extend the ideas from the proof of Lemma C.8. The variation $\Upsilon^{t,\tau}$ satisfies the differential equation

$$\begin{aligned}\frac{\mathrm{d}}{\mathrm{d}t}\Upsilon^{t,\tau} &= A(x_2(t))\,\Psi_{x_2}(t, \tau) - A(x_1(t))\,\Psi_{x_1}(t, \tau)\\ &= A(x_2(t))\,\Upsilon^{t,\tau} + \big[A(x_2(t)) - A(x_1(t))\big]\Psi_{x_1}(t, \tau),\end{aligned}$$

which leads to a variation of constants integral that can be estimated as

$$\begin{aligned}\big\|\Upsilon^{t,\tau}\big\| &\le \int_\tau^t \big\|\Psi_{x_2}(t, \sigma)\big\|\,\|A(x_2(\sigma)) - A(x_1(\sigma))\|\,\big\|\Psi_{x_1}(\sigma, \tau)\big\|\,\mathrm{d}\sigma\\ &\le \int_\tau^t C_Y\, e^{\rho_Y(t-\sigma)}\, \varepsilon_{\tilde{A}}(d_\rho(x_2, x_1))\, e^{\mu\,\sigma}\, C_Y\, e^{\rho_Y(\sigma-\tau)}\mathrm{d}\sigma\\ &\le C_Y^2\, e^{\rho_Y(t-\tau)}\, \varepsilon_{\tilde{A}}(d_\rho(x_2, x_1))\frac{e^{\mu\,\tau}}{-\mu}.\end{aligned} \tag{3.57}$$

Here we use ideas from Appendix B; we applied Corollary B.3 to obtain A as a uniformly continuous fiber mapping $\mathcal{B}^\rho_\beta(I; X) \to B^\mu(I; \mathcal{L}(Y))$ with continuity modulus $\varepsilon_{\tilde{A}}$ (that depends on μ). □

Thus, the flow $\Psi_x^{t,\tau}$ depends uniformly continuously on $x \in \mathcal{B}^\rho_\beta(I; X)$ when viewed as a flow with ρ_Y-exponential growth *and* measured with an additional exponential factor $e^{\mu\,\tau}$.

Remark 3.33 In the previous proposition, if A is α-Hölder continuous, then we can replace μ by $\alpha\,\rho$ to obtain a similar, α-Hölder continuous result using Lemma B.2. ◊

To show that $x \mapsto \underset{\sim}{\mathrm{D}}_x\Psi_x(t, \tau)$ is continuous as well, we first write down the corresponding variation in the bundle trivialization chart:

$$
\begin{aligned}
&\big(\underset{\sim}{\mathrm{D}}_x \Psi_{x_2} \cdot \widetilde{\delta x} - \underset{\sim}{\mathrm{D}}_x \Psi_{x_1} \cdot \widetilde{\delta x}\big)(t,\tau) \\
&\quad = \int_\tau^t \Psi_{x_2}(t,\sigma)\big[\mathrm{D}A(x_2(\sigma))\,\Pi(x_2|_0^\sigma)\,\Pi(\gamma_{x_2(0),\bar{x}})\,\widetilde{\delta x}(\sigma)\big]\,\Psi_{x_2}(t,\sigma) \\
&\qquad - (2 \rightsquigarrow 1)\mathrm{d}\sigma \\
&\quad = \int_\tau^t \Psi_{x_2}(t,\sigma)\big[\mathrm{D}A(x_2(\sigma))_e\,\widetilde{\delta x}(\sigma)\big]\,\Psi_{x_2}(t,\sigma) - (2 \rightsquigarrow 1)\mathrm{d}\sigma,
\end{aligned}
\tag{3.58}
$$

where the notation $(2 \rightsquigarrow 1)$ means that we take the first expression and replace all 2's by 1's (note that $\gamma_{x_2(0),\bar{x}} = \gamma_0$ in the first term and $\Pi(\gamma_{x_1(0),\bar{x}}) = \mathbb{1}$ in the second term). The last line is just a rewrite in terms of the frame as in (3.55) and suppresses all parallel transport terms. We separately estimate continuity of the three factors in the integrand, and insert the estimate of Proposition 3.32 for the variation $\Psi_\bullet$ in the first and third factor. Note that $\widetilde{\delta x}$ is the same over both curves x_1 and x_2 in this trivialization.

For the middle factor $\mathrm{D}A(x(t))_e$, we again apply Nemytskii operator techniques from Appendix B. But in this case we have to combine these with holonomy terms, due to the fact that comparison of $\mathrm{D}A$ at nearby points $\xi_2,\ \xi_1 \in X$ only makes sense after identification of the tangent spaces $\mathrm{T}_{\xi_2}X$ and $\mathrm{T}_{\xi_1}X$.

Proposition 3.34 *Let $A \in C^1_{b,u}$ according to Definition 2.9. Then for any $\mu < 0$, the map*

$$
x \mapsto \big(t \mapsto \mathrm{D}A(x(t))_e\big) : U_{\bar{x}} \subset \mathcal{B}^\rho_\beta(I;X) \to B^\mu\big(I; \mathcal{L}(\mathrm{T}_{\bar{x}}X; \mathcal{L}(Y))\big) \tag{3.59}
$$

is uniformly continuous. If moreover $A \in C^{1,\alpha}_{b,u}$, then the map (3.59) is α-Hölder with μ replaced by $\alpha\,\rho$.

Proof Let $x_1,\ x_2 \in U_{\bar{x}}$. We introduce another frame f to directly compare $\mathrm{D}A$ at points $x_1(t),\ x_2(t)$. Let $f_{x_1(t)} = e_{x_1(t)} : \mathbb{R}^n \to \mathrm{T}_{x_1(t)}X$ and define $f_{x_2(t)} = \Pi(\gamma_t) \cdot f_{x_1(t)}$. Thus, the frames e and f at $x_2(t)$ are both defined in terms of the frame $e_{\bar{x}}$; $e_{x_2(t)}$ by parallel transport along $x_2 \circ \gamma_0$ and $f_{x_2(t)}$ by parallel transport along $\gamma_t \circ x_1$, see Fig. 3.7. Since $f_{x_1(t)} = e_{x_1(t)}$, we can rewrite the difference of (3.59) at points on these curves as

$$
\begin{aligned}
\mathrm{D}A(x_2(t))_e - \mathrm{D}A(x_1(t))_e = &\big[\mathrm{D}A(x_2(t))_e - \mathrm{D}A(x_2(t))_f\big] \\
&+ \big[\mathrm{D}A(x_2(t))_f - \mathrm{D}A(x_1(t))_f\big].
\end{aligned}
$$

The first term can be estimated by the holonomy defect along the loop

$$
\gamma_0^{-1} \circ x_2|_t^0 \circ \gamma_t \circ x_1|_0^t
$$

using Lemma 2.19 and the second term using the continuity of $\mathrm{D}A$ and Proposition 2.13. Together, this leads to

$$
\begin{aligned}
&\|\mathrm{D}A(x_2(t))_e - \mathrm{D}A(x_1(t))_e\| \\
&\quad \le \|\mathrm{D}A\|\,\|\mathbb{1} - \Pi\big(\gamma_0^{-1} \circ x_2|_t^0 \circ \gamma_t \circ x_1|_0^t\big)\| + \varepsilon_{\mathrm{D}A}\big(d(x_2(t), x_1(t))\big) \\
&\quad \le C_v\, C\, d_\rho(x_2, x_1)\, e^{\rho\, t} + \varepsilon_{\mathrm{D}A}\big(d(x_2(t), x_1(t))\big).
\end{aligned}
$$

If $d_\rho(x_2, x_1)\, e^{\rho\, t} \ge \delta_X$, then we use the boundedness estimate $\|\mathbb{1} - \Pi(\gamma)\| \le 2$ for any closed loop γ and Remark 2.12 to effectively extend the local to a global continuity modulus. We can recover any α-Hölder continuity from the Lipschitz holonomy estimate, again by using the fact that the holonomy is bounded by 2 in combination with Lemma 1.20.

With the same arguments as in Lemma B.2, it follows that $x \mapsto \big(t \mapsto \mathrm{D}A(x(t))_e\big)$ is uniformly or α-Hölder continuous, and we denote its continuity modulus by $\varepsilon_{\widetilde{\mathrm{D}A}}$. Note that $\varepsilon_{\widetilde{\mathrm{D}A}}$ does not depend on the trivialization chart since all estimates are uniform with respect to these charts. □

Proposition 3.35 *For any $\mu < 0$ and uniformly in $\bar{x} \in X$, the map $x \mapsto \underset{\sim}{\mathrm{D}}_x \Psi_x(t, \tau)$ satisfies continuity estimate (3.60) in a trivialization neighborhood $U_{\bar{x}} \subset \mathcal{B}^\rho_\beta(I; X)$.*

Proof We combine the estimates from Propositions 3.32 and 3.34 and obtain for (3.58)

$$
\begin{aligned}
&\|\big(\underset{\sim}{\mathrm{D}}_x \Psi_{x_2} \cdot \widetilde{\delta x} - \underset{\sim}{\mathrm{D}}_x \Psi_{x_1} \cdot \widetilde{\delta x}\big)(t, \tau)\| \\
&\le \int_\tau^t \|\Psi_{x_2}(t, \sigma)\, \big[\mathrm{D}A(x_2(\sigma))_e\, \widetilde{\delta x}(\sigma)\big]\, \Psi_{x_2}(\sigma, \tau) - (2 \rightsquigarrow 1)\| \mathrm{d}\sigma \\
&\le \int_\tau^t \Big(\|\Psi_{x_2}(t, \sigma) - \Psi_{x_1}(t, \sigma)\|\, \|\mathrm{D}A(x_2(\sigma))_e\|\, \big\|\Psi_{x_2}(\sigma, \tau)\big\| \\
&\quad + \big\|\Psi_{x_1}(t, \sigma)\big\|\, \|\mathrm{D}A(x_2(\sigma))_e - \mathrm{D}A(x_1(\sigma))_e\|\, \big\|\Psi_{x_2}(\sigma, \tau)\big\| \\
&\quad + \big\|\Psi_{x_1}(t, \sigma)\big\|\, \|\mathrm{D}A(x_1(\sigma))_e\|\, \|\Psi_{x_2}(\sigma, \tau) - \Psi_{x_1}(\sigma, \tau)\|\Big) \big\|\widetilde{\delta x}\big\|_\rho\, e^{\rho\,\sigma} \mathrm{d}\sigma \\
&\le \int_\tau^t \Big(C_Y^2\, e^{\rho_Y (t-\sigma)}\, \varepsilon_A(d_\rho(x_2, x_1))\, \frac{e^{\mu\,\sigma}}{-\mu}\, C_v\, C_Y\, e^{\rho_Y(\sigma-\tau)} \\
&\quad + C_Y\, e^{\rho_Y(t-\sigma)}\, \varepsilon_{\widetilde{\mathrm{D}A}}(d_\rho(x_2, x_1))\, e^{\mu\,\sigma}\, C_Y\, e^{\rho_Y(\sigma-\tau)} \\
&\quad + C_Y\, e^{\rho_Y(t-\sigma)}\, C_v\, C_Y^2\, e^{\rho_Y(\sigma-\tau)}\, \varepsilon_A(d_\rho(x_2, x_1))\, \frac{e^{\mu\,\tau}}{-\mu}\Big) \big\|\widetilde{\delta x}\big\|_\rho\, e^{\rho\,\sigma} \mathrm{d}\sigma \\
&\le C\, e^{\rho_Y(t-\tau)}\, \varepsilon(d_\rho(x_2, x_1))\, e^{(\rho+\mu)\tau}\, \big\|\widetilde{\delta x}\big\|_\rho.
\end{aligned}
\tag{3.60}
$$

We absorbed all constants and integration factors such as $\frac{1}{-\mu}$ into the general constant C and combine the continuity moduli into one; C_v is a global bound on all vector fields including A and its derivatives. □

We finally plug estimate (3.60) into Eq. (3.43c). We repeat the Nemytskii and holonomy arguments for $\tilde{f}$ and $\mathrm{D}_x \tilde{f}$ (just as for A and $\mathrm{D}A$) to obtain a uniform continuity estimate for

$$x \mapsto \underset{\sim}{\mathrm{D}}_x T_Y(x, y) : U_{\tilde{x}} \to \mathcal{L}\big(B^{\rho}(I; \mathrm{T}_{\tilde{x}} X); B^{\rho+\mu}(I; Y)\big).$$

both for any $\mu < 0$, or with $\mu = \alpha\,\rho$ when $A,\ f \in C^{1,\alpha}_{b,u}$. That is, $\underset{\sim}{\mathrm{D}}_x \mathrm{T}_Y(\,\cdot\,, y)$ is a map that given a curve $x \in \mathcal{B}^{\rho}_{\beta}(I; X)|_{C^1}$, linearly maps a variational curve δx over x to a variational curve δy in the trivial bundle $\underset{\sim}{\mathrm{T}} B^{\rho}_{\eta}(I; Y)$. We can formulate this more abstractly as

$$\underset{\sim}{\mathrm{D}}_x \mathrm{T}_Y(\,\cdot\,, y) \in \Gamma^{\alpha}_{b,u}\big(\mathcal{B}^{\rho}_{\beta}(I; X)|_{C^1}; \mathcal{L}\big(\underset{\sim}{\mathrm{T}}\mathcal{B}^{\rho}_{\beta}(I; X)\big|_{C^1}; B^{\rho+\mu}(I; Y)\big)\big),$$

that is, $\underset{\sim}{\mathrm{D}}_x \mathrm{T}_Y(\,\cdot\,, y)$ is a uniformly α-Hölder bounded section of the bounded geometry bundle

$$\pi : \mathcal{L}\big(\underset{\sim}{\mathrm{T}}\mathcal{B}^{\rho}_{\beta}(I; X)\big|_{C^1}; B^{\rho+\mu}(I; Y)\big) \to \mathcal{B}^{\rho}_{\beta}(I; X)|_{C^1}.$$

Continuity in the Other Cases

We treated the continuity for one of the maps (**??**) with respect to a single variable. The continuity in all other cases can be shown in a similar fashion. Many arguments can be repeated, but each of these maps also has its own peculiar details which makes that I have not been able to find one general, abstract way to prove continuity of all of these maps at once. In this section we shall focus on these specific details and not repeat the recurring elements. Let me reiterate that the uniform continuity results hold for any $\mu < 0$ sufficiently small, and these can be replaced by α-Hölder continuity when μ is replaced by $\alpha\,\rho$ and the spectral gap condition (1.11) is satisfied for $r = 1 + \alpha$.

First of all, note that continuity with respect to the combined variables follows directly from continuity with respect to each separate variable since we have explicit uniform or Hölder continuity moduli. If $f(x, y)$ has continuity moduli $\varepsilon_x,\ \varepsilon_y$ with respect to $x,\ y$, respectively, then

$$\begin{aligned}\|f(x_2, y_2) - f(x_1, y_1)\| &\leq \|f(x_2, y_2) - f(x_1, y_2)\| + \|f(x_1, y_2) - f(x_1, y_1)\| \\ &\leq \varepsilon_x(d(x_2, x_1)) + \varepsilon_y(d(y_2, y_1)) \\ &\leq (\varepsilon_x + \varepsilon_y)\big(d((x_2, y_2), (x_1, y_1))\big)\end{aligned}$$

shows that $\varepsilon_x + \varepsilon_y$ is a continuity modulus for f. We assumed w.l.o.g. that $\varepsilon_x,\ \varepsilon_y$ are non-decreasing, while all choices of distance on the product space are equivalent, so we leave it unspecified.

Let us start with the easy cases. Continuity of the map (3.43c) as a function of y, that is,

$$y \mapsto \underset{\sim}{\mathrm{D}}_x T_Y(x, y) : B^{\rho}_{\eta}(I; Y) \to \mathcal{L}\big(B^{\rho}(I; x^*(\mathrm{T}X)); B^{\rho+\mu}(I; Y)\big),$$

requires no additional details: only f and $\mathrm{D}_x f$ depend on $y \in Y$, and we can reapply the arguments above to show that these depend continuously on $y \in B^\rho_\eta(I;Y)$. No holonomy terms are present since $\underset{\sim}{\mathrm{T}}B^\rho_\eta(I;Y)$ is a trivial bundle. That is, we can directly compare $\underset{\sim}{\mathrm{D}}_x T_Y$ at different $y_1,\ y_2 \in B^\rho_\eta(I;Y)$; keeping $x \in \mathcal{B}^\rho_\beta(I;X)$ fixed means that everything is situated in the fixed fiber $\underset{\sim}{\mathrm{T}}_x\mathcal{B}^\rho_\beta(I;X)|_{C^1} = B^\rho(I;x^*(\mathrm{T}X))$ and no holonomy terms are required.

Continuity of the map (3.43d), i.e. $\underset{\sim}{\mathrm{D}}_y T_Y(x,\ y)$, both with respect to x and y follows along the same lines. Neither case requires holonomy arguments; we just apply the Nemytskii technique to $\mathrm{D}_y f$ and reuse Proposition 3.32 to show continuity with respect to x.

The formal derivatives (3.43a) and (3.43b) of T_X map into $\underset{\sim}{\mathrm{T}}\mathcal{B}^\rho_\beta(I;X)\big|_{C^1}$; here we have to apply holonomy arguments in the codomain. Let us first focus on

$$x_0 \mapsto \underset{\sim}{\mathrm{D}}_{x_0} T_X(y,x_0) : X \to \mathcal{L}\big(\mathrm{T}_{x_0}X;\ B^{\rho+\mu}(I;\mathrm{T}_{\bar{x}}X)\big)$$

with a local trivialization $U_{\bar{x}} \times B^{\rho+\mu}(I;\mathrm{T}_{\bar{x}}X)$ within[13] the bundle $\underset{\sim}{\mathrm{T}}\mathcal{B}^{\rho+\mu}_\beta(I;X)\big|_{C^1}$ with additional μ in the exponential growth norm on the fibers. Note that $\underset{\sim}{\mathrm{D}}_{x_0} T_X(y,\ \cdot)$ could actually be considered as a bundle map on the vector bundle $\mathrm{T}X$ that is linear on each tangent space $\mathrm{T}_{x_0}X$. We consider a local trivialization of $\mathrm{T}B(\bar{x};\delta_X) \subset \mathrm{T}X$ by parallel transport along geodesics: this is equivalent to trivialization by a normal coordinate chart for the purpose of measuring continuity, while it matches the trivialization of $\underset{\sim}{\mathrm{T}}\mathcal{B}^\rho_\beta(I;X)|_{U_{\bar{x}}}$. This will lead to a holonomy term.

For any $x_0 \in B(\bar{x};\delta_X)$ we have $T_X(y,x_0) \in U_{\bar{x}}$ by construction, so let e denote the frame introduced by the trivialization of $\underset{\sim}{\mathrm{T}}\mathcal{B}^\rho_\beta(I;X)|_{U_{\bar{x}}}$, i.e. by parallel transport along solution curves $T_X(y,x_0)$. On the other hand, let f denote a frame introduced by local parallel transport. We define

$$f_{x_1(t)} = e_{x_1(t)} \quad \text{and} \quad f_{x_2(t)} = \Pi(\gamma_t) \cdot f_{x_1(t)}.$$

It follows from Lemma C.10 that $\underset{\sim}{\mathrm{D}}_{x_0} T_X(y,\ \cdot) = \big(t \mapsto \mathrm{D}\Phi_y(t,0,\ \cdot)\big)$ satisfies the correct type of continuity estimates, but with respect to local charts (or equivalently, with respect to f determined by local parallel transport) instead of the choice of frame e, defined by the topology of $\underset{\sim}{\mathrm{T}}\mathcal{B}^\rho_\beta(I;X)\big|_{C^1}$. To examine the difference, let $x_{0,1},\ x_{0,2} \in B(\bar{x};\delta_X)$ denote two initial conditions and $x_i = T_X(y,x_{0,i}),\ i = 1,2$, their respective solution curves for a fixed $y \in B^\rho_\eta(I;Y)$. We also fix $x_{0,1} = \bar{x}$ for convenience. Then we have

[13] Embeddings $B^\rho \hookrightarrow B^{\rho+\mu}$ are continuous, so we can view $U_{\bar{x}} \times B^{\rho+\mu}(I;\mathrm{T}_{\bar{x}}X)$ as a local trivialization of a subset of $\underset{\sim}{\mathrm{T}}\mathcal{B}^{\rho'}_\beta(I;X)$ with $\rho' = \rho + \mu$.

$$\begin{aligned}
&\mathrm{D}\Phi_y(t,0,x_{0,2})_e - \mathrm{D}\Phi_y(t,0,x_{0,1})_e \\
&\quad = \big[\mathrm{D}\Phi_y(t,0,x_{0,2})_e - \mathrm{D}\Phi_y(t,0,x_{0,2})_f\big] \\
&\qquad + \big[\mathrm{D}\Phi_y(t,0,x_{0,2})_f - \mathrm{D}\Phi_y(t,0,x_{0,1})_f\big] \\
&\quad = \big[\Pi(\gamma_0^{-1}\circ x_2|_t^0) - \Pi(x_1|_t^0\circ\gamma_t^{-1})\big]\cdot \mathrm{D}\Phi_y(t,0,x_{0,2})\cdot\Pi(\gamma_0) \\
&\qquad + \big[\mathrm{D}\Phi_y(t,0,x_{0,2})_f - \mathrm{D}\Phi_y(t,0,x_{0,1})_f\big].
\end{aligned}$$

This shows that uniform and Hölder continuity with respect to the topology of $\underset{\sim}{\mathrm{T}}\mathcal{B}^\rho_\beta(I;X)\big|_{C^1}$ is equivalent to the same continuity with respect to normal coordinate charts, since the additional holonomy term can be estimated in the same way as in Proposition 3.34. Continuity of

$$y \mapsto \underset{\sim}{\mathrm{D}}_{x_0} T_X(y,x_0) : B^\rho_\eta(I;Y) \to \mathcal{L}\big(\mathrm{T}_{x_0}X; B^{\rho+\mu}(I;\mathrm{T}_{\tilde{x}}X)\big)$$

follows in the same way, if we first apply Corollary C.12 to obtain the continuity estimates with respect to the frame f.

Finally, we consider continuity of the map (3.43b),

$$\underset{\sim}{\mathrm{D}}_y T_X(y,x_0) \in \mathcal{L}\big(B^\rho(I;Y); B^{\rho+\mu}(I;\mathrm{T}_{\tilde{x}}X)\big)$$

with respect to $y \in B^\rho_\eta(I;Y)$ and $x_0 \in X$. We apply Corollary C.12 and Lemma 3.29 to conclude that $\mathrm{D}\Phi_y(t,\tau,x_y(\tau))$ depends α-Hölder or uniformly continuously on y. Lemma 3.29 in combination with a Nemytskii operator argument shows that $\mathrm{D}_y\tilde{v}_X$ induces a uniformly continuous map

$$y \mapsto \big(t \mapsto \mathrm{D}_y\tilde{v}_X(x_y(t),y(t))\big) : B^\rho_\eta(I;Y) \to B^\mu\big(I;\mathcal{L}(Y;\mathrm{T}X)\big)$$

with μ replaced by $\alpha\,\rho$ in the Hölder case. For continuity with respect to $x_0 \in X$ we need to replace application of Corollary C.12 by that of Lemma C.10 for dependence of $x_y = T_X(y,x_0)$ on x_0. Again the continuity estimates obtained are with respect to the frame e and we use Lemma 2.19 to estimate the additional holonomy term when switching to the frame f.

3.7.6 Application of the Fiber Contraction Theorem

In Sect. 3.7.3 we already established that the fiber mapping $(T,\ \underset{\sim}{D}T)$ in formula (3.46) is uniformly contractive. With the results of the previous sections on formal tangent bundles and continuous formal derivatives, we can now apply the fiber contraction theorem, see Appendix D.

Proposition 3.36 *For any $\mu < 0$, the fiber mapping (3.46) has a unique, globally attractive fixed point $(\Theta^\infty, \underset{\sim}{D}\Theta^\infty) \in \mathcal{S}_0 \times \mathcal{S}_1^\mu$, while it also holds that $\underset{\sim}{D}\Theta^\infty \in \mathcal{S}_1^0$.*

Proof In the notation of Theorem D.1 we take $X = \mathcal{S}_0$ and $Y = \mathcal{S}_1^\mu$ as in (3.45) with ρ, μ such that $\rho_Y < \rho + \mu < \rho < \rho_X$ holds. The fiber mapping is $F = (T, \underset{\sim}{D}T)$, as in (3.46). The first two conditions of Theorem D.1 are satisfied due to the arguments in Sect. 3.7.3, while the third condition that $\underset{\sim}{D}T$ is continuous can be obtained from the results in Sect. 3.7.5 as follows.

First, note that $(T, \underset{\sim}{D}T)$ is a well-defined, uniformly contractive fiber mapping both when acting on $B^\rho(I; Y)$ and on $B^{\rho+\mu}(I; Y)$ variational curves. Thus, for each $n \geq 0$ we have $\underset{\sim}{D}\Theta^n \in \mathcal{S}_1^0 \hookrightarrow \mathcal{S}_1^\mu$, where the embedding is continuous. The same conclusion holds for $\underset{\sim}{D}\Theta^\infty$ by a simple uniform contraction argument. Next, we view $\underset{\sim}{D}T$ as a map

$$\underset{\sim}{D}T : \mathcal{S}_0 \times \mathcal{S}_1^0 \to \mathcal{S}_1^\mu. \tag{3.61}$$

Note that we set $\mu = 0$ in the domain only. To obtain continuity of (3.61) with respect to the base variable $\Theta \in \mathcal{S}_0$, it is sufficient to check that the maps

$$\begin{aligned} y &\mapsto \underset{\sim}{D}_y T(y, x_0) : B_\eta^\rho(I; Y) \to \mathcal{L}\big(B^\rho(I; Y); B^{\rho+\mu}(I; Y)\big), \\ y &\mapsto \underset{\sim}{D}_{x_0} T(y, x_0) : B_\eta^\rho(I; Y) \to \mathcal{L}\big(\mathrm{T}_{x_0}X; B^{\rho+\mu}(I; Y)\big) \end{aligned} \tag{3.62}$$

are uniformly continuous, uniformly in $x_0 \in X$. Continuity of (3.61) with respect to the base $\mathcal{S}_0$ (with fixed fiber part $\underset{\sim}{D}\Theta \in \mathcal{S}_1^0 \hookrightarrow \mathcal{S}_1^\mu$) then follows from the interpretation of (3.62) as acting on maps $(\Theta, \underset{\sim}{D}\Theta)$ with the supremum norm on $\mathcal{S}_0$. The maps (3.62) are defined by the chain rule formula (3.44) in terms of the derivative maps (**??**). A variation of $y \in B_\eta^\rho(I; Y)$ can be distributed over the product (we only estimate the variation of $\underset{\sim}{D}_y T$ with respect to y, but the variation of $\underset{\sim}{D}_{x_0} T$ is completely analogous),

$$\begin{aligned} &\|\underset{\sim}{D}_y T(y_2, x_0) - \underset{\sim}{D}_y T(y_1, x_0)\|_{\rho+\mu,\rho} \\ &\quad\leq \|\underset{\sim}{D}_x T_Y(x_{y_2}, y_2) \cdot \big[\underset{\sim}{D}_y T_X(y_2, x_0) - \underset{\sim}{D}_y T_X(y_1, x_0)\big]\|_{\rho+\mu,\rho} \\ &\qquad + \|\big[\underset{\sim}{D}_x T_Y(x_{y_2}, y_2) - \underset{\sim}{D}_x T_Y(x_{y_1}, y_1)\big] \cdot \underset{\sim}{D}_y T_X(y_1, x_0)\|_{\rho+\mu,\rho} \\ &\quad\leq \|\underset{\sim}{D}_x T_Y(x_{y_2}, y_2)\|_{\rho+\mu,\rho+\mu} \cdot \|\underset{\sim}{D}_y T_X(y_2, x_0) - \underset{\sim}{D}_y T_X(y_1, x_0)\|_{\rho+\mu,\rho} \\ &\qquad + \|\underset{\sim}{D}_x T_Y(x_{y_2}, y_2) - \underset{\sim}{D}_x T_Y(x_{y_1}, y_1)\|_{\rho+\mu,\rho} \cdot \|\underset{\sim}{D}_y T_X(y_1, x_0)\|_{\rho,\rho}. \end{aligned} \tag{3.63}$$

The $\|\cdot\|_{\rho_2,\rho_1}$ denote operator norms on linear (bundle) maps from B^{ρ_1} to B^{ρ_2} spaces. In the factor that is not varied we can simply take the operator norm between functions of either ρ or $\rho + \mu$ exponential growth: in Sect. 3.7.3 we have seen that the fiber maps are uniformly bounded linear in both cases. The factor that is varied satisfies a uniform continuity estimate in $\|\cdot\|_{\rho+\mu,\rho}$-norm, a result from Sect. 3.7.5. Note that we use the topology defined in Sect. 3.7.4 on the intermediate space $\underset{\sim}{\mathrm{T}}\mathcal{B}^{\rho}_{\beta}(I;X)\big|_{C^1}$, as well as a local trivialization to express the difference $\underset{\sim}{\mathrm{D}}_y T_X(y_2,x_0) - \underset{\sim}{\mathrm{D}}_y T_X(y_1,x_0)$.

As a result of the fiber contraction theorem, we conclude that there is a unique, globally attractive fixed point $(\Theta^\infty, \underset{\sim}{\mathrm{D}}\Theta^\infty)$ of the fiber mapping (3.46). Note that $\underset{\sim}{\mathrm{D}}\Theta^\infty$ is already well-defined as an element of $\mathcal{S}^0_1$, although it is only proven to be attractive in $\mathcal{S}^\mu_1$. □

As a next step, we show that the fixed point map $\underset{\sim}{\mathrm{D}}\Theta^\infty$ that we found is actually continuous. This follows from a standard uniform contraction argument.

Proposition 3.37 *For any $\mu < 0$, the map $\underset{\sim}{\mathrm{D}}\Theta^\infty \in \mathcal{S}^\mu_1$ is uniformly continuous. If we set $\mu \le \alpha\,\rho$ and the assumptions of Theorem 3.2 are satisfied with $r \ge 1+\alpha$, then it is α-Hölder continuous.*

Proof First note that it is sufficient to prove the statement for $\mu < 0$ sufficiently small, or $\mu = \alpha\,\rho$ in case of α-Hölder continuity; by continuous embedding of exponential growth spaces, it then automatically follows for any μ that is more negative. We use local trivializations by parallel transport to express continuity moduli of functions with domain $\mathrm{T}X$.

The assumptions of Theorem 3.2 imply that the spectral gap condition $\rho_Y < \rho + \mu < \rho < \rho_X$ is satisfied. Since $\underset{\sim}{\mathrm{D}}\Theta^\infty$ is (the fiber part of) the fixed point of the uniform contraction $(T, \underset{\sim}{\mathrm{D}}T)$, we have for any two $x_1,\ x_2 \in B(\bar{x};\delta_X) \subset X$ that

$$
\begin{aligned}
&\|\underset{\sim}{\mathrm{D}}\Theta^\infty(x_2) - \underset{\sim}{\mathrm{D}}\Theta^\infty(x_1)\|_{\rho+\mu} \\
&\quad = \|\underset{\sim}{\mathrm{D}}_y T(\Theta^\infty(x_2),x_2)\cdot \underset{\sim}{\mathrm{D}}\Theta^\infty(x_2) + \underset{\sim}{\mathrm{D}}_{x_0} T(\Theta^\infty(x_2),x_2) - (2 \rightsquigarrow 1)\|_{\rho+\mu} \\
&\quad \le \|\underset{\sim}{\mathrm{D}}_y T(\Theta^\infty(x_2),x_2) - \underset{\sim}{\mathrm{D}}_y T(\Theta^\infty(x_1),x_1)\|_{\rho+\mu,\rho}\cdot \|\underset{\sim}{\mathrm{D}}\Theta^\infty(x_2)\|_{\rho} \\
&\qquad + \|\underset{\sim}{\mathrm{D}}_y T(\Theta^\infty(x_1),x_1)\|_{\rho+\mu,\rho+\mu}\cdot \|\underset{\sim}{\mathrm{D}}\Theta^\infty(x_2) - \underset{\sim}{\mathrm{D}}\Theta^\infty(x_1)\|_{\rho+\mu} \\
&\qquad + \|\underset{\sim}{\mathrm{D}}_{x_0} T(\Theta^\infty(x_2),x_2) - \underset{\sim}{\mathrm{D}}_{x_0} T(\Theta^\infty(x_1),x_1)\|_{\rho+\mu} \\
&\quad \le \varepsilon_{\underset{\sim}{\mathrm{D}}_y T}\big((L+1)d(x_2,x_1)\big) + q\,\|\underset{\sim}{\mathrm{D}}\Theta^\infty(x_2) - \underset{\sim}{\mathrm{D}}\Theta^\infty(x_2)\|_{\rho+\mu} \\
&\qquad + \varepsilon_{\underset{\sim}{\mathrm{D}}_x T}\big((L+1)d(x_2,x_1)\big).
\end{aligned}
$$

Here $L = \mathrm{Lip}(\Theta^\infty)$ denotes the Lipschitz constant of $\Theta^\infty \in \mathcal{S}_0$, while $q < 1$ is the uniform contraction factor of $\underset{\sim}{\mathrm{D}}_y T$ on the fibers of $B^{\rho}_{\eta}(I;Y) \times B^{\rho+\mu}(I;Y)$. We saw in Sect. 3.7.5 that the maps $\underset{\sim}{\mathrm{D}}_y T,\ \underset{\sim}{\mathrm{D}}_x T$ have appropriate continuity moduli into $B^{\rho+\mu}(I;Y)$. Finally, we move the contraction term to the left-hand side, divide by $1-q$, and obtain

$$\|\underset{\sim}{\mathrm{D}}\Theta^{\infty}(x_2) - \underset{\sim}{\mathrm{D}}\Theta^{\infty}(x_1)\|_{\rho+\mu} \leq \frac{1}{1-q}\Big[\varepsilon_{\underset{\sim}{\mathrm{D}}_y T}\big((L+1)d(x_2, x_1)\big) + \varepsilon_{\underset{\sim}{\mathrm{D}}_x T}\big((L+1)d(x_2, x_1)\big)\Big].$$

This shows that $\underset{\sim}{\mathrm{D}}\Theta^{\infty} \in \mathcal{S}_1^{\mu}$ has the same type of continuity modulus as $\underset{\sim}{\mathrm{D}}T$. □

3.7.7 Derivatives on Banach Manifolds

We can recover the maps (**??**) as true derivatives on Banach manifolds if we restrict to bounded time intervals $J \subset I = \mathbb{R}_{\leq 0}$. The maps T_X, T_Y naturally restrict to such intervals, either exactly, or in a well-behaved approximate way. By restricting to intervals $J = [a, 0]$ with $a < 0$, the spaces $B_\eta^\rho(J; Y)$ and $\mathcal{B}_\beta^\rho(J; X)$ become Banach manifolds and the restrictions of T_X, T_Y become continuously differentiable maps on these.

Lemma 3.38 *For any $-\infty < a < 0$, the spaces $B_\eta^\rho(J; Y)$ and $\mathcal{B}_\beta^\rho(J; X)$ with $J = [a, 0]$ a bounded interval are well-defined Banach manifolds.*

Proof We first treat the easy case $B_\eta^\rho(J; Y)$. For any $-\infty < a < 0$, the norms $\|\cdot\|_\rho$ and $\|\cdot\|_0$ are equivalent on $B^\rho(J; Y)$. The set $B_\eta^\rho(J; Y)$ is an open ball of radius η in the Banach space $B^0(J; Y)$, so it follows that $B_\eta^\rho(J; Y)$ is a Banach manifold as an open subset of $B^\rho(J; Y)$.

In the same way, the metrics d_ρ and d_0 are equivalent on $\mathcal{B}^\rho(J; X)$, but here we need to do a little more work to show the following. □

Proposition 3.39 *The set $\mathcal{B}_\beta^\rho(J; X)$ is open in $\mathcal{B}^\rho(J; X)$.*

Proof Let $x \in \mathcal{B}_\beta^\rho(J; X)$, hence by Definition 3.19, x is approximated on each interval of length $\|[t_1, t_2]\| \leq T$ by $t \mapsto \Phi(t, t_2, x(t_2))$, where Φ denotes the flow of $v_X \circ g$, the horizontal part of the unperturbed vector field (3.10). The map

$$(t, t_2) \mapsto d\big(x(t), \Phi(t, t_2, x(t_2))\big)$$

is continuous, and since it is defined on a compact subset of $J \times J$, it attains its supremum

$$\eta_1 = \sup_{t_2 \in J} \sup_{t \in [t_2 - T, t_2]} d\big(x(t), \Phi(t, t_2, x(t_2))\big),$$

so it must hold that $\eta_1 < \beta$. Let $\tilde{x} \in B(x; \eta_2) \subset \mathcal{B}^\rho(J; X)$ with

$$\eta_2 = (\beta - \eta_1)\frac{e^{-\rho a}}{1 + C_X\, e^{\rho_X T}}.$$

We apply the triangle inequality and obtain

$$\begin{aligned}
d\big(\tilde{x}(t), \Phi(t, t_2, \tilde{x}(t_2))\big) &\le d\big(\tilde{x}(t), x(t)\big) + d\big(x(t), \Phi(t, t_2, x(t_2))\big) \\
&\quad + d\big(\Phi(t, t_2, x(t_2)), \Phi(t, t_2, \tilde{x}(t_2))\big) \\
&\le e^{\rho\, a}\, \eta_2 + \eta_1 + C_X\, e^{\rho_X\, T}\, e^{\rho\, a}\, \eta_2 \\
&\le \big(1 + C_X\, e^{\rho_X\, T}\big) e^{\rho\, a}\, \eta_2 + \eta_1 < \beta.
\end{aligned}$$

This shows that all functions in the ball $B(x; \eta_2) \subset \mathcal{B}^\rho(J; X)$ are still (β, T)-approximate solutions of $v_X \circ g$, and thus $\mathcal{B}^\rho_\beta(J; X)$ is open. □

From here on we shall not always precisely distinguish between $\mathcal{B}^\rho_\beta(J; X)$ and $\mathcal{B}^\rho(J; X)$ anymore.

We introduce a local coordinate chart κ_x around a curve $x \in \mathcal{B}^\rho(J; X)$ using the exponential map (see also [Kli95, Sect. 2.3]):

$$\kappa_x : U_x \subset \mathcal{B}^\rho(J; X) \to B^\rho(J; x^*(\mathrm{T}X)) : \xi \mapsto \big(t \mapsto \exp^{-1}_{x(t)}(\xi(t))\big). \tag{3.64}$$

The vector bundle $x^*(\mathrm{T}X)$ is trivial, so the space of sections $B^\rho(J; x^*(\mathrm{T}X))$ is isomorphic to $B^\rho(J; \mathbb{R}^n)$. An explicit trivialization of $x^*(\mathrm{T}X)$ (and thus isomorphism of sections) can be obtained, for example if $x \in C^1$, using parallel transport as in (3.50) and identification of $\mathrm{T}_{x(0)}X \cong \mathbb{R}^n$ by a choice frame, but we refrain from making such a choice here; one reason is that curves $x \in \mathcal{B}^\rho(J; X)$ are only assumed continuous. The chart κ_x bijectively covers a full $r_{\mathrm{inj}}(X)$ neighborhood of x with respect to the metric d_0, hence a neighborhood of size $r_{\mathrm{inj}}(X)\, e^{-\rho\, a} > 0$ with respect to d_ρ. Recall that δ_X is X-small as in Definition 2.8; let us choose a radius $\delta_a = \delta_X\, e^{-\rho\, a}$, such that all bounded geometry results also hold true in these induced charts κ_x. Then the coordinate transition map

$$\kappa_{x_2} \circ \kappa_{x_1}^{-1} : B^\rho(J; x_1^*(\mathrm{T}X)) \to B^\rho(J; x_2^*(\mathrm{T}X)) \tag{3.65}$$

is a bijection between isomorphic Banach spaces that is as smooth as the exponential map of X. □

Remark 3.40 We could choose isomorphisms $\tau_x : B^\rho(J; x^*(\mathrm{T}X)) \to B^\rho(J; \mathbb{R}^n)$ to obtain one fixed Banach space $B^\rho(J; \mathbb{R}^n)$ as model for the manifold $\mathcal{B}^\rho(J; X)$. The τ_x are linear isometries so they preserve norms and smoothness, hence there is no need to explicitly make this identification. Specifically, note that the construction of $B^\rho(J; x^*(\mathrm{T}X))$ as the pullback along a curve x that is merely continuous, does not influence the smoothness of coordinate transformations on $\mathcal{B}^\rho(J; X)$. ◊

We shall again call charts in this atlas 'normal coordinate charts', since they are induced by normal coordinates on X along the curve $x \in \mathcal{B}^\rho(J; X)$. By construction all bounded geometry results carry over to these induced charts. In particular, we can measure maps in terms of their coordinate representations. We will use this fact without always explicitly mentioning it.

The tangent space of $\mathcal{B}^{\rho}(J; X)$ at a point x can be canonically identified as

$$\mathrm{T}_x \mathcal{B}^{\rho}(J; X) \cong B^{\rho}(J; x^*(\mathrm{T}X)) \tag{3.66}$$

as follows. Let $s \mapsto x_s : (-\varepsilon, \varepsilon) \subset \mathbb{R} \to \mathcal{B}^{\rho}(J; X)$ be a C^1 family of curves such that $x_0 = x$ and let $\xi_s = \kappa_x(x_s)$ be their representation in the coordinate chart $B^{\rho}(J; x^*(\mathrm{T}X))$. The chart κ_x is induced by normal coordinates, so

$$d_{\rho}(x_0, x_s) = \sup_{t \in J} d(x_0(t), x_s(t))\, e^{\rho t} = \sup_{t \in J} \| \exp_{x_0(t)}^{-1}(x_s(t)) \|\, e^{\rho t} = \|\xi_s\|_{\rho}$$

shows that $\|\cdot\|_{\rho}$ is the canonical norm on the chart $B^{\rho}(J; x^*(\mathrm{T}X))$. Then $v = \frac{\mathrm{d}}{\mathrm{d}s}\xi_s\big|_{s=0}$ represents a tangent vector in $\mathrm{T}_x\mathcal{B}^{\rho}(J; X)$, while $v \in B^{\rho}(J; x^*(\mathrm{T}X))$ by construction.

This completes our exposition of the manifold structure of $\mathcal{B}^{\rho}(J; X)$. We shall again exclusively make use of induced normal coordinate charts (3.64), in order to use results on bounded geometry.

The map T_X can be restricted to curves on any subinterval $J = [a, 0] \subset I = \mathbb{R}_{\leq 0}$. Let us introduce the restriction operator on curves

$$\rho_a : C(I; Z) \to C(J; Z) : z \mapsto z|_J. \tag{3.67}$$

This operator acts naturally on $B^{\rho}_{\eta}(I; Y)$ and $\mathcal{B}^{\rho}_{\beta}(I; X)$ and there is a natural family of restrictions T_X^a of T_X such that

$$T_X^a \circ \rho_a = \rho_a \circ T_X \qquad \text{for any } -\infty < a < 0. \tag{3.68}$$

Proposition 3.41 *Let $J = [a, 0]$ with $-\infty < a < 0$. Then*

$$T_X^a : B^{\rho}_{\eta}(J; Y) \times X \to \mathcal{B}^{\rho}_{\beta}(J; X) \tag{3.69}$$

is a differentiable map between Banach manifolds with partial derivatives given by a natural restriction of the maps (3.43a) and (3.43b).

Proof Let $s \mapsto y + s\,\delta y \in B^{\rho}_{\eta}(J; Y)$ be a one-parameter family of curves and let

$$\kappa_x : B(x; \delta_a) \subset \mathcal{B}^{\rho}_{\beta}(J; X) \to B^{\rho}(J; x^*(\mathrm{T}X))$$

be an induced normal coordinate chart centered around the curve $x = T_X^a(y, x_0)$. The map T_X is Lipschitz, so for s sufficiently small, $T_X^a(y + s\,\delta y, x_0)$ maps into $B(x; \delta_a)$. The vector field $\tilde{v}_X(\,\cdot\,, (y + s\,\delta y)(t))$ depends smoothly on the parameter s and generates $x_s = T_X^a(y + s\,\delta y, x_0)$. We apply Theorem E.2 with

$$\frac{\mathrm{d}}{\mathrm{d}s}\Big[\tilde{v}_X\big(x(t),(y+s\,\delta y)(t)\big)\Big]_{s=0} = \mathrm{D}_y\tilde{v}_X(x(t),y(t))\cdot\delta y(t)$$

to obtain (3.43b) as the pointwise derivative of $\mathrm{ev}_t \circ T_X^a$, for any $t \in J$.

Now we only need to show that (3.43b) viewed as derivative pointwise in t satisfies linear approximation estimates, uniformly for all $t \in J$ with respect to d_ρ. We work in the local chart κ_X, so $x_s(t)$ is represented in the normal coordinate chart centered at $x(t)$, while the curve $s \mapsto y + s\,\delta y$ is canonically represented in $B^\rho(I;Y)$ with derivative δy.

Since $s \mapsto \big(\kappa_X \circ T_X^a(y+s\,\delta y, x_0)\big)(t) \in C^1(\mathbb{R}; \mathrm{T}_{x(t)}X)$, we can apply the mean value theorem to estimate

$$\begin{aligned}
&\|\big[T_X^a(y+s\,\delta y,x_0) - T_X^a(y,x_0) - \underset{\sim}{\mathrm{D}}_y T_X(y,x_0)\cdot s\,\delta y\big](t)\| \\
&\quad\le \|\big[\big(\underset{\sim}{\mathrm{D}}_y T_X(y+\sigma_t\,\delta y,x_0) - \underset{\sim}{\mathrm{D}}_y T_X(y,x_0)\big)\cdot s\,\delta y\big](t)\| \qquad (3.70)\\
&\quad\le \|\underset{\sim}{\mathrm{D}}_y T_X(y+\sigma_t\,\delta y,x_0) - \underset{\sim}{\mathrm{D}}_y T_X(y,x_0)\|\;\|\delta y\|_\rho\, e^{\rho t}\,|s|
\end{aligned}$$

for some $\sigma_t \in (0,s)$. Note that σ_t will in general depend on $t \in J$, so there is (a priori) not one curve $y + \sigma\,\delta y$ such that (3.70) holds for all $t \in J$ at once. In Sect. 3.7.5 we showed that $\underset{\sim}{\mathrm{D}}_y T_X : \underset{\sim}{\mathrm{T}}B_\eta^\rho(I;Y) \to \underset{\sim}{\mathrm{T}}\mathcal{B}^{\rho+\mu}(I;X)|_{C^1}$ is continuous; on the bounded interval J the norms $\|\cdot\|_\rho$ and $\|\cdot\|_{\rho+\mu}$ are equivalent, so $\underset{\sim}{\mathrm{D}}_y T_X$ is continuous into $B^\rho(J;x^*(\mathrm{T}X))$ as well. Using this fact, we plug the result above into the definition of (directional) derivative and verify

$$\begin{aligned}
&\lim_{s\to 0}\frac{1}{s}\,\|T_X^a(y+s\,\delta y,x_0) - T_X^a(y,x_0) - \underset{\sim}{\mathrm{D}}_y T_X(y,x_0)\cdot s\,\delta y\|_\rho \\
&\le \lim_{s\to 0}\sup_{t\in J}\frac{|s|}{s}\,\|\underset{\sim}{\mathrm{D}}_y T_X(y+\sigma_t\,\delta y,x_0) - \underset{\sim}{\mathrm{D}}_y T_X(y,x_0)\|\;\|\delta y\|_\rho = 0.
\end{aligned}$$

Therefore, the derivative of T_X^a at (y,x_0) in the direction of δy is given by $\underset{\sim}{\mathrm{D}}_y T_X(y,x_0)\cdot\delta y$ restricted to the interval J. The limit is uniform on $\|\delta y\|_\rho = 1$ and this map is continuous and linear in δy, so T_X^a is continuously partially differentiable with respect to y.

If we use a local chart around $x_0 \in X$, then we find in the same way that T_X^a is continuously partially differentiable with respect to x_0. Thus, T_X^a is (continuously) differentiable. □

The map T_Y does not have a similarly natural restriction since it depends on the complete 'history' of the curves x, y through the integral from $-\infty$. The dependence on earlier times is exponentially suppressed, though. Therefore, we construct a family of restrictions that approach T_Y when the amount of additional history in the input goes to infinity. Let $-\infty < a \le b < 0$ and define the family $\mathrm{T}_Y^{b,a}$ of restrictions as

$$\mathrm{T}_Y^{b,a} : \mathcal{B}^\rho_\beta([a,0]; X) \times B^\rho_\eta([a,0]; Y) \to B^\rho_\eta([b,0]; Y),$$
$$(x, y) \mapsto \Big(t \mapsto \int_a^t \Psi_x(t,\tau)\, \tilde{f}(x(\tau), y(\tau))\mathrm{d}\tau\Big) \quad \text{for each } t \in [b,0]. \tag{3.71}$$

Proposition 3.42 *The family $T_Y^{b,a}$ approximates T_Y in the sense that for any fixed $b \in (-\infty, 0]$, we have*

$$T_Y^{b,a} \circ \rho_a \to \rho_b \circ T_Y \tag{3.72}$$

when $a \to -\infty$, uniformly in $x, y \in \mathcal{B}^\rho_\beta(I; X) \times B^\rho_\eta(I; Y)$.

Proof This follows from straightforward estimates:

$$\begin{aligned}\|T_Y^{b,a} \circ \rho_a(x,y) - \rho_b \circ T_Y(x,y)\|_\rho &\le \sup_{t\in[b,0]} e^{-\rho t} \int_{-\infty}^a \|\Psi_x(t,\tau)\, \tilde{f}(x(\tau), y(\tau))\|\mathrm{d}\tau \\ &\le \sup_{t\in[b,0]} e^{-\rho t} \int_{-\infty}^a C_Y\, e^{\rho_Y(t-\tau)}\, \zeta\, \mathrm{d}\tau \\ &\le \frac{C_Y\, \zeta}{-\rho_Y}\, e^{\rho_Y(b-a)-\rho b}.\end{aligned}$$

□

In a same way we define approximate families for (3.43c) and (3.43d), denoted by $\underset{\sim}{\mathrm{D}}_x T_Y^{b,a}$ and $\underset{\sim}{\mathrm{D}}_y T_Y^{b,a}$, respectively.

Corollary 3.43 *The families $\underset{\sim}{\mathrm{D}}_x T_Y^{b,a}$ and $\underset{\sim}{\mathrm{D}}_y T_Y^{b,a}$ approximate (3.43c) and (3.43d) in the same way as in Proposition 3.42.*

Proposition 3.44 *Let $-\infty < a \le b < 0$. Then $T_Y^{b,a}$ is a differentiable map between Banach manifolds.*

Proof We shall only show that $T_Y^{b,a}$ is continuously partially differentiable respect to x. Continuous partial differentiability with respect to y follows along the same lines and total differentiability then is a direct consequence of these (also in the Banach manifold setting, see [Lan95, Prop. 3.5]).

Let $x \in \mathcal{B}^\rho_\beta(J; X)$ and $y \in B^\rho_\eta(J; Y)$ with $J = [a, 0]$. Let κ_x be an induced normal coordinate chart around x and let

$$x_s = x + s\,\delta x \in B^\rho(J; x^*(\mathrm{T}X))$$

be a one-parameter family of curves in $\mathcal{B}^\rho_\beta(J; X)$, represented in the chart κ_x (for s sufficiently small). Then $\delta x \in B^\rho(J; x^*(\mathrm{T}X))$ is naturally identified as the derivative $\frac{\mathrm{d}}{\mathrm{d}s} x_s\big|_{s=0}$.

We shall show that the partial derivative $\mathrm{D}_x T_Y^{b,a}$ is given by the formal derivative (3.43c), but with J as domain of integration and interpreted as a mapping into $B^\rho([b,0]; Y)$. Again, we split the full expression into manageable pieces and apply the mean value theorem.

$$
\begin{aligned}
&\| [T_Y^{b,a}(x_s, y) - T_Y^{b,a}(x, y) - \mathrm{D}_x T_Y^{b,a}(x, y) \cdot s\,\delta x](t) \| \\
&\le \int_a^t \| \Psi_{x_s}(t,\tau)\,\tilde f(x_s(\tau), y(\tau)) - \Psi_x(t,\tau)\,\tilde f(x(\tau), y(\tau)) \\
&\quad - \Psi_x(t,\tau)\,\mathrm{D}_x \tilde f(x(\tau), y(\tau))\, s\delta x(\tau) - (\underset{\sim}{\mathrm{D}}_x \Psi_x \cdot s\delta x)(t,\tau)\tilde f(x(\tau), y(\tau)) \| \mathrm{d}\tau \\
&\le \int_a^t \| \Psi_{x_s}(t,\tau) - \Psi_x(t,\tau) - (\underset{\sim}{\mathrm{D}}_x \Psi_x \cdot s\,\delta x)(t,\tau) \|\, \| \tilde f(x(\tau), y(\tau)) \| \\
&\quad + \| \Psi_x(t,\tau) \|\, \| \tilde f(x_s(\tau), y(\tau)) - \tilde f(x(\tau), y(\tau)) - \mathrm{D}_x \tilde f(x(\tau), y(\tau))\, s\,\delta x(\tau) \| \\
&\quad + \| \Psi_{x_s}(t,\tau) - \Psi_x(t,\tau) \|\, \| \tilde f(x_s(\tau), y(\tau)) - \tilde f(x(\tau), y(\tau)) \| \mathrm{d}\tau
\end{aligned}
$$

and application of Theorem E.2 shows that formula (3.43e) for $\underset{\sim}{\mathrm{D}}_x \Psi_{x_s} \cdot \delta x$ is the derivative of Ψ_{x_s}. We use this for a mean value theorem estimate[14] in the first and third[15] term to arrive at

$$
\begin{aligned}
&\le \int_a^t \| (\underset{\sim}{\mathrm{D}}_x \Psi_{x_\sigma} \cdot s\,\delta x)(t,\tau) - (\underset{\sim}{\mathrm{D}}_x \Psi_x \cdot s\,\delta x)(t,\tau) \|\, \zeta \\
&\quad + C_Y\, e^{\rho_Y (t-\tau)}\, \| \mathrm{D}_x \tilde f(x_\sigma(\tau), y(\tau)) - \mathrm{D}_x \tilde f(x(\tau), y(\tau)) \|\, |s|\, \| \delta x \|_\rho\, e^{\rho\,\tau} \\
&\quad + \| (\underset{\sim}{\mathrm{D}}_x \Psi_{x_\sigma} \cdot s\,\delta x)(t,\tau) \|\, \| \mathrm{D}_x \tilde f(x_\sigma(\tau), y(\tau)) \|\, |s|\, \| \delta x \|_\rho\, e^{\rho\,\tau} \mathrm{d}\tau \\
&\le \int_a^t C\, e^{\rho_Y (t-\tau)}\, \varepsilon(|\sigma|\, \| \delta x \|_\rho)\, e^{(\rho+\mu)\tau}\, |s|\, \| \delta x \|_\rho\, \zeta \\
&\quad + C_Y\, e^{\rho_Y (t-\tau)}\, \varepsilon_{\mathrm{D}_x f}\big(|\sigma|\, \| \delta x \|_\rho\, e^{\rho\,\tau} \big)\, |s|\, \| \delta x \|_\rho\, e^{\rho\,\tau} \\
&\quad + \frac{C_Y^2\, C_v}{-\rho}\, \| \delta x \|_\rho\, e^{\rho_Y (t-\tau)}\, e^{\rho\, t}\, \zeta\, |s|\, \| \delta x \|_\rho\, e^{\rho\,\tau} \mathrm{d}\tau.
\end{aligned}
$$

We applied Proposition 3.35 to estimate the variation of $\underset{\sim}{\mathrm{D}}_x \Psi$; the induced normal coordinate charts and Proposition 2.13 allow us to freely switch between parallel transport and normal coordinates for estimating differences. All exponential norms are equivalent on the compact interval J, so with the usual estimates we see that this expression is $o(|s|)$, uniformly for all $\| \delta x \|_\rho = 1$. □

We have thus converted the map $T = \mathrm{T}_Y \circ (T_X, \mathrm{pr}_1)$ to a Banach manifold setting by defining it on curves restricted to compact time intervals. Although all estimates

---

[14] The intermediate point σ in the mean value theorem implicitly depends on both t and τ and will be different in each term. This does not affect the uniform estimates, so we suppress this dependence in the notation.

[15] We applied the intermediate value theorem to both factors in the third term. This is not strictly necessary: we could also have applied it to only one of these, and apply a uniform continuity estimate to the other term. That would still have yielded a size estimate $\varepsilon(|s|)\,|s| = o(|s|)$. When we generalize to higher derivatives, we shall make use of this fact: at least one of the factors will be differentiable and yield a factor $|s|$, while the other term(s) can be estimated by a continuity modulus $\varepsilon(|s|)$.

were already in place, this technicality allows us to draw the conclusions of the final points *6* and *7* in the scheme in Sect. 3.7.1.

Lemma 3.45 (The Θ^n have true derivatives) *Fix $\mu < 0$ and let $\Theta^n : X \to B^\rho_\eta(I; Y)$ be differentiable into $B^{\rho+\mu}(I; Y)$. Recursively define $\Theta^{n+1}(x_0) = T(\Theta^n(x_0), x_0)$. Then Θ^{n+1} is again differentiable into $B^{\rho+\mu}(I; Y)$.*

Proof We define $\mathrm{D}\Theta^{n+1} \in \mathcal{S}^0_1$ using (3.46) and proceed to show that it is the derivative of Θ^{n+1} as a function $\mathrm{D}\Theta^{n+1} \in \mathcal{S}^\mu_1$ by a direct estimate

$$\|\Theta^{n+1}(x_0 + h) - \Theta^{n+1}(x_0) - \mathrm{D}\Theta^{n+1}(x_0) \cdot h\|_{\rho+\mu} \le \varepsilon \, \|h\|,$$

with $x_0,\ x_0 + h \in X$ represented in normal coordinate charts.

First, we use the Nemytskii operator technique to get rid of the infinite tail $t \to -\infty$. For any given $\varepsilon > 0$, we have on $(-\infty, b]$ the crude estimate

$$\begin{aligned}
&\sup_{t \le b} \|\big[\Theta^{n+1}(x_0 + h) - \Theta^{n+1}(x_0) - \mathrm{D}\Theta^{n+1}(x_0) \cdot h\big](t)\| \, e^{-(\rho+\mu)\,t} \\
&\quad \le \Big(\|\Theta^{n+1}(x_0 + h) - \Theta^{n+1}(x_0)\|_\rho + \|\mathrm{D}\Theta^{n+1}(x_0) \cdot h\|_\rho\Big)\, e^{-\mu\, b} \\
&\quad \le \big(\mathrm{Lip}(\Theta^{n+1}) + \|\mathrm{D}\Theta^{n+1}\|\big)\, \|h\| \, e^{-\mu\, b} \le \varepsilon \, \|h\|
\end{aligned}$$

for some $b(\varepsilon)$ that is sufficiently negative. We use the differentiability of $T^{b,a} = T^{b,a}_Y \circ (T^a_X, \mathrm{pr}_1)$ on the finite interval $[b, 0]$ that is left. We define $\Theta^{n+1}_{b,a} = T^{b,a} \circ \rho_a \circ \Theta^n$ and estimate

$$\begin{aligned}
&\sup_{t \in [b,0]} \|\big[\Theta^{n+1}(x_0 + h) - \Theta^{n+1}(x_0) - \mathrm{D}\Theta^{n+1}(x_0) \cdot h\big](t)\| \, e^{-(\rho+\mu)\,t} \\
&\quad \le \|\rho_b \circ \Theta^{n+1}(x_0 + h) - \Theta^{n+1}_{b,a}(x_0 + h)\|_{\rho+\mu} \\
&\quad\quad + \|\rho_b \circ \Theta^{n+1}(x_0) - \Theta^{n+1}_{b,a}(x_0)\|_{\rho+\mu} \\
&\quad\quad + \|\big[\rho_b \circ \mathrm{D}\Theta^{n+1}(x_0) - \mathrm{D}\Theta^{n+1}_{b,a}(x_0)\big] \cdot h\|_{\rho+\mu} \\
&\quad\quad + \|\Theta^{n+1}_{b,a}(x_0 + h) - \Theta^{n+1}_{b,a}(x_0) - \mathrm{D}\Theta^{n+1}_{b,a}(x_0) \cdot h\|_{\rho+\mu}.
\end{aligned}$$

This holds for all $a \le b$ and the first three terms can be made arbitrarily small when $a \to -\infty$ due to Proposition 3.42 and Corollary 3.43, while the last term is $o(\|h\|)$ since $\Theta^{n+1}_{b,a}$ is differentiable by the chain rule. If the estimate $o(\|h\|)$ is independent of a, then we can finally, for any $\varepsilon > 0$ and $\|h\| \le \delta$ sufficiently small, estimate this by $\varepsilon \, \|h\|$.

That the term $o(\|h\|)$ is independent of a follows from another application of the mean value theorem:

$$
\begin{aligned}
&\|\Theta^{n+1}_{b,a}(x_0+h) - \Theta^{n+1}_{b,a}(x_0) - \mathrm{D}\Theta^{n+1}_{b,a}(x_0)\cdot h\|_{\rho+\mu}\\
&\quad\le \|\mathrm{D}\Theta^{n+1}_{b,a}(\xi) - \mathrm{D}\Theta^{n+1}_{b,a}(x_0)\|_{\rho+\mu}\,\|h\|\\
&\qquad\qquad\qquad\qquad \text{where } d(\xi,x_0)\le\|h\|\\
&\quad= \|\mathrm{D}T^{b,a}(\Theta^n(\xi),\xi)\cdot\rho_a\circ\mathrm{D}\Theta^n(\xi)\\
&\qquad - \mathrm{D}T^{b,a}(\Theta^n(x_0),x_0)\cdot\rho_a\circ\mathrm{D}\Theta^n(x_0)\|_{\rho+\mu}\,\|h\|\\
&\quad\le \Big(\|\mathrm{D}T^{b,a}(\Theta^n(\xi),\xi) - \mathrm{D}T^{b,a}(\Theta^n(x_0),x_0)\|_{\rho+\mu,\rho}\,\big\|\mathrm{D}\Theta^n(x_0)\big\|_{\rho}\\
&\qquad + \|\mathrm{D}T^{b,a}(\Theta^n(x_0),x_0)\|_{\rho+\mu,\rho+\mu}\,\big\|\mathrm{D}\Theta^n(\xi)-\mathrm{D}\Theta^n(x_0)\big\|_{\rho+\mu}\Big)\,\|h\|\\
&\quad\le \varepsilon(d(\xi,x_0))\,\|h\|
\end{aligned}
$$

since the continuity estimates for the formal derivatives $\underset{\sim}{\mathrm{D}}T$ directly translate into the same estimates for the true derivative counterparts $\mathrm{D}T^{b,a}$ on restricted intervals. □

Thus, we can now conclude by induction, starting at $\Theta^0 \equiv 0$, that for each $n \ge 0$ the map $\Theta^n \in \mathcal{S}_0$ is differentiable when viewed as map into $B^{\rho+\mu}(I;Y)$, while the results in Sect. 3.7.5 show that we actually have

$$\Theta^n \in C^{1,\alpha}_{b,u}\big(X;\, B^{\rho+\mu}(I;Y)\big). \tag{3.73}$$

Finally, we have uniformly convergent sequences

$$\Theta^n \to \Theta^\infty \in \mathcal{S}_0 \quad\text{and}\quad \mathrm{D}\Theta^n \to \underset{\sim}{\mathrm{D}}\Theta^\infty \in \mathcal{S}_1^\mu \tag{3.74}$$

by the fiber contraction theorem, so now we apply Theorem D.2 (taking into account Remark D.3) to conclude that $\underset{\sim}{\mathrm{D}}\Theta^\infty$ is the derivative of Θ^∞ as a map $X \to B^{\rho+\mu}(I;Y)$. It was already shown in Proposition 3.37 that $\underset{\sim}{\mathrm{D}}\Theta^\infty$ is bounded and continuous, just as the $\mathrm{D}\Theta^n$ in (3.73).

Remark 3.46 (*on topologies used*) The convergence in (3.74) is with respect to uniform supremum norms as in Definition 2.9. These induce a topology that is stronger than the weak Whitney (or compact-open) topology, cf. Sect. 1.7. The convergence in Theorem D.2 is with respect to the weak Whitney topology, both the assumption and result. This is sufficient, since we are primarily interested in the result that Θ^∞ is differentiable, not in what sense Θ^n and its derivatives converge to Θ^∞. On the other hand, we did already have convergence of $\mathrm{D}\Theta^n \to \underset{\sim}{\mathrm{D}}\Theta^\infty = \mathrm{D}\Theta^\infty$ with respect to these stronger uniform norms, so clearly $\Theta^n \to \Theta^\infty$ in uniform C^1-norm as well. ◊

3.7.8 Conclusion for the First Derivative

The evaluation map $\mathrm{ev}_0 : B^{\rho+\mu}(I; Y) \to Y$ is bounded linear so the graph (3.41) of the persistent invariant manifold also satisfies

$$\tilde{h} = \mathrm{ev}_0 \circ \Theta^\infty \in C^{1,\alpha}_{b,u}(X; Y).$$

The size of $\mathrm{D}\tilde{h}$ can be estimated using the fixed point equation for $\underset{\sim}{\mathrm{D}}\Theta^\infty$. This yields

$$\|\underset{\sim}{\mathrm{D}}\Theta^\infty\|_\rho \leq \frac{q}{1-q} \|\underset{\sim}{\mathrm{D}}_{x_0} T_X\|,$$

and the contraction factor $q < 1$ can be made arbitrarily small by choosing ζ small. As indicated in (3.18), ζ is in turn controlled by $\delta,\ \sigma_1$ from Theorem 3.2, and ν from Lemma 3.12, which can be chosen arbitrarily small. This completes the proof of all statements in Theorem 3.2 for $r = 1 + \alpha$ with $\alpha \in [0, 1]$. Note that this is the case $k = 1$ as in Remark 3.3, 5.

3.7.9 Higher Order Derivatives

To obtain higher order smoothness of the perturbed invariant manifold, we consider Eq. (3.42) for $k > 1$. The principal term governing the contraction is still $\underset{\sim}{\mathrm{D}}_y T(\Theta(x_0), x_0)$, now acting on multilinear maps $\underset{\sim}{\mathrm{D}}^k\Theta(x_0) \in \mathcal{L}^k\big(\mathrm{T}X; B^{k\rho+\mu_k}(I; Y)\big)$. The remaining terms only depend on lower order derivatives of $\Theta(x_0)$, hence they do not influence the contractivity estimate in the fiber contraction theorem. It must be verified, though, that these terms depend continuously on the lower order derivatives as mappings into $B^{k\rho+\mu_k}(I; Y)$. Note again that we set $\mu_k = \alpha\,\rho$ in case of α-Hölder continuity; in case of uniform continuity (denoted by $\alpha = 0$) we choose a sequence $\{\mu_j\}_{1\leq j\leq k}$ such that the following hold true:

1. $\mu_j < 0$ for each j;
2. the spectral gap condition $\rho_Y < k\,\rho + \mu_k < \rho < \rho_X$ still holds;
3. there exists a $\tilde{\rho} < \rho$ such that

$$k\,\rho + \mu_k < k\,\tilde{\rho} \quad \text{and} \quad j\,\tilde{\rho} \leq j\,\rho + \mu_j \quad \text{for any } j < k. \tag{3.75}$$

It follows that the sequence μ_j's is strictly decreasing (i.e. increasing in absolute value), and that we have continuous embeddings $B^{k\tilde{\rho}} \hookrightarrow B^{k\rho+\mu_k}$ and $B^{j\rho+\mu_j} \hookrightarrow B^{j\tilde{\rho}}$; in the first embedding we reserved some spectral space to apply Corollary B.3. These choices—as well as more ideas in this section—are inspired by [Van89, Sect. 3], which is an interesting read for comparison in a simpler setting.

We reuse the scheme already defined in Sect. 3.7.1 for the first order derivatives (**??**). Let us walk through these items step by step and indicate the changes that need to be made.

1. Candidate functions for the higher order derivatives can be found by formal differentiation and application of Theorem E.2. This is a straightforward procedure, although tedious and quite unenlightening to perform. Let us show just one example[16]:

$$\begin{aligned}
&\underset{\sim}{\mathrm{D}}_y^2 T_X(y, x_0)(\delta y_1, \delta y_2)(t) \\
&= \int_t^0 \Big(\mathrm{D}\Phi_y(t, \tau, x_y(\tau)) \cdot \Big[\mathrm{D}_y^2 \tilde{v}_X(x_y(\tau), y(\tau))\big(\delta y_1(\tau), \delta y_2(\tau)\big) \\
&\quad + \mathrm{D}_x \mathrm{D}_y \tilde{v}_X(x_y(\tau), y(\tau))\big(\delta y_1(\tau), \big(\underset{\sim}{\mathrm{D}}_y T_X(y, x_0)\delta y_2\big)(\tau)\big)\Big] \\
&\quad + \Big[\mathrm{D}^2\Phi_y(t, \tau, x_y(\tau)) \cdot \big(\underset{\sim}{\mathrm{D}}_y T_X(y, x_0)\delta y_2\big)(\tau) \\
&\quad + \int_t^\tau \mathrm{D}\Phi_y(t, \sigma, x_y(\sigma)) \cdot \big[\mathrm{D}_y \mathrm{D}_x \tilde{v}_X(x_y(\sigma), y(\sigma)) \cdot \delta y_2(\sigma)\big] \cdot \mathrm{D}\Phi_y(\sigma, \tau, x_y(\tau)) \mathrm{d}\sigma \Big] \\
&\quad \cdot \mathrm{D}_y \tilde{v}_X(x_y(\tau), y(\tau)) \cdot \delta y_1(\tau) \Big) \mathrm{d}\tau.
\end{aligned} \tag{3.76}$$

2. For contractivity in the fibers we still only need to consider the map $\underset{\sim}{\mathrm{D}}_y T$ as in (3.44), since that is the principal term in (3.42). This map is contractive for any $\rho' \in (\rho_Y, \rho_X)$, hence also for $\rho' = k\,\rho + \mu$ for any $\mu \leq 0$ sufficiently small, when $\rho \in (\rho_Y, \rho_X)$ is chosen appropriately. The other terms in (3.42) are bounded maps as well, and linear in the $\underset{\sim}{\mathrm{D}}^j \Theta(x_0)$. It follows from Proposition C.3 that each of these terms has weighted degree

$$\sum_{i=1}^{k-1} i \cdot p_i = k - m$$

with respect to the $\underset{\sim}{\mathrm{D}}^j \Theta(x_0)$, while they incur an additional exponential factor $e^{m\,\rho\,t}$ from taking m derivatives with respect to $x_0 \in X$, due to Lemma C.1. Thus the combined exponential growth rates sum to $k\,\rho$, and $\rho_Y < k\,\rho$ implies that the variation of constants integrals still converge, so these terms are bounded maps into $B^{k\rho}$ spaces. This still holds if we add μ_j's that satisfy the conditions set out above.

In the notation of Appendix C we define spaces of higher order derivatives,

$$\mathcal{S}_k^\mu = \Gamma_b\big(\mathcal{L}^k(\mathrm{T}X; B^{k\rho+\mu}(I; Y))\big) \tag{3.77}$$

---

[16] Even though it is not obvious from (3.76), this expression is in fact symmetric in δy_1, δy_2. To verify this for the terms containing $\mathrm{D}_x \mathrm{D}_y \tilde{v}_X$, one should change the order of integration of τ, σ and expand the expression $\big(\underset{\sim}{\mathrm{D}}_y T_X(y, x_0)\delta y_2\big)(\tau)$ using (3.43b).

with norms

$$\|\underset{\sim}{\mathrm{D}}^k\Theta\| = \sup_{x_0\in X} \|\underset{\sim}{\mathrm{D}}^k\Theta(x_0)\|_{\mathcal{L}^k(\mathrm{T}_{x_0}X;\, B^{k\rho+\mu}(I;Y))}$$

extending (3.45). Similarly, we define as extensions of (3.46), higher order fiber mappings

$$F^{(k)} = (T,\ \underset{\sim}{\mathrm{D}}^1 T,\ \ldots,\ \underset{\sim}{\mathrm{D}}^k T) \quad \text{on} \quad \mathcal{S}_0 \times \mathcal{S}_1^{\mu_1} \times \cdots \times \mathcal{S}_k^{\mu_k}. \tag{3.78}$$

These are again uniform fiber contractions with respect to the final factor $\mathcal{S}_k^{\mu_k}$ as fiber, for any choice of $\mu_j \le 0$ sufficiently small.

3. Instead of trying to construct higher order formal tangent bundles, we represent the higher derivatives on 'formal tensor bundles'

$$\underset{\sim}{\mathrm{T}}\mathcal{B}^{\rho}_{\beta}(I;X)^k = \coprod_{x\in\mathcal{B}^{\rho}_{\beta}(I;X)} B^{\rho}(I; x^*(\mathrm{T}X))^{\otimes k}. \tag{3.79}$$

Note that this choice of representation along base curves x matches our choice to represent higher derivatives as in Definition C.6 when the former is evaluated at a fixed t. The trivializations, then, are defined by tensor products of parallel transport terms $\Pi(x|_0^t)^{\otimes k}$, again when restricted to curves $x \in C^1$. The resulting holonomy terms can be estimated by either the $k$þpower of the single holonomy term, or, $k-1$ factors can be bounded by $\|\Pi(\gamma)\| \le 2$ such that the remaining factor fulfills the required α-Hölder estimate. Then, all details in Sect. 3.7.5 can be repeated to obtain uniform or α-Hölder continuity of the higher derivatives of T_X, T_Y as maps on these formal tensor bundles. Note that we can break each expression into parts such that only one factor is varied for the continuity estimate and thus only once adds either $\alpha\,\rho$ or μ to the exponential growth rate. Thus, the spectral gap condition is still satisfied.

4. We apply the fiber contraction theorem to (3.78) with base $\mathcal{S}_0 \times \mathcal{S}_1^{\mu_1} \times \cdots \times \mathcal{S}_{k-1}^{\mu_{k-1}}$ and fiber $\mathcal{S}_k^{\mu_k}$. In case of $\alpha = 0$, we again seize some of the unused spectral space for the carefully chosen μ_j's, such that condition *3* of Theorem D.1 holds. Let us assume by induction that $F^{(k-1)}$ already is a globally attractive fiber map. The conditions (3.75) imply that if we insert elements $\underset{\sim}{\mathrm{D}}^j\Theta \in \mathcal{S}_j^{\mu_j}$ into $F^{(k)}$, then their exponents sum at most to $k\,\tilde{\rho}$, so the mapping onto the fiber $\mathcal{S}_k^{\mu_k}$ is continuous by application of Corollary B.3. For α-Hölder continuity we can simply choose $\mu_j = \alpha\,\rho$ for all $1 \le j \le k$. Thus, we find a globally attractive fixed point

$$(\Theta^\infty,\ \underset{\sim}{\mathrm{D}}\Theta^\infty, \ldots,\ \underset{\sim}{\mathrm{D}}^k\Theta^\infty) \in \mathcal{S}_0 \times \mathcal{S}_1^{\mu_1} \times \cdot \times \mathcal{S}_k^{\mu_k} \quad \text{with} \quad \underset{\sim}{\mathrm{D}}^k\Theta^\infty \in C^{\alpha}_{b,u}.$$

5. We constructed a manifold structure on $\mathcal{B}^{\rho}(J;X)$ (and a trivial one on $B^{\rho}_{\eta}(J;Y)$ as well) with an atlas of charts induced by normal coordinate charts of the underlying

manifold X. We represent higher[17] derivatives in these induced normal coordinate charts $B^\rho(J; x^*(\mathrm{T}X))$. Thus, we have for example

$$\mathrm{D}_x^k T_X^a(x, y) \in \mathcal{L}^k\big(B^\rho(J; x^*(\mathrm{T}X)); B^\rho(J; Y)\big).$$

This precisely matches the representation of the formal higher derivatives on the tensor space $B^\rho(I; x^*(\mathrm{T}X))^{\otimes k}$ in point 3. Higher differentiability of the restricted maps T_X^a and $T_Y^{b,a}$ follows as in Sect. 3.7.7.

6. Lemma 3.45 can be generalized to prove by induction over n that higher derivatives $\mathrm{D}^k\Theta^n$ exist; we define $\mathrm{D}^k\Theta^{n+1} \in \mathcal{S}_k^0$ by (3.78).
7. By induction we may assume that it was already proven that

$$\Theta^n \to \Theta^\infty \in C_{b,u}^{k-1}\big(X; B^{(k-1)\rho+\mu_{k-1}}(I; Y)\big) \quad \text{as } n \to \infty.$$

We apply Corollary D.4 to conclude that $\Theta^n \to \Theta^\infty$ as sequence of C^k functions and that $\Theta^\infty \in C_{b,u}^k\big(X; B^{k\rho+\mu_k}(I; Y)\big)$. The convergence is with respect to our uniform supremum norms, see Remark 3.46.

Just as in Sect. 3.7.8, this finalizes the proof of all statements in Theorem 3.2, but now for $r = k + \alpha$ with $k > 1$ and $\alpha \in [0, 1]$. It follows that $\tilde{h} \in C_{b,u}^{k,\alpha}$, but the last part that remains to be shown, though, is that $\|\tilde{h}\|_{k-1}$ can be made as small as desired.

From the fixed point Eq. (3.42) it follows that

$$\|\mathrm{D}^{k-1}\Theta^\infty(x_0)\| \le \frac{1}{1-q} \sum_{\substack{l,m\ge 0 \\ l+m\le k-1 \\ (l,m)\ne(0,0),(1,0)}} \|\mathrm{D}_y^l \mathrm{D}_{x_0}^m T(\Theta^\infty(x_0), x_0) \cdot P_{l,k-m}\big(\mathrm{D}^\bullet\Theta^\infty(x_0)\big)\|.$$

Each term with $l \ge 1$ contains at least one factor $\mathrm{D}^j\Theta^\infty(x_0)$ with $j < k-1$; these can be assumed to be small by induction. The one remaining term with $(l, m) = (0, k-1)$ can be expanded using Proposition C.3. This yields, suppressing arguments $\Theta^\infty(x_0)$ and x_0,

$$\mathrm{D}_{x_0}^{k-1} T = \sum_{j=1}^{k-1} \mathrm{D}_x^j T_Y \cdot P_{j,k-1}(\mathrm{D}_{x_0}^\bullet T_X).$$

Since all terms are uniformly bounded in appropriate norms, it suffices to show that the $\mathrm{D}_x^j T_Y$ can be made small. Recall formula (3.43c) and the fiber contraction estimate for $\mathrm{D}_x T_Y$ in Sect. 3.7.3, where we saw that $\mathrm{D}_x T_Y$ could be made small by choosing $\|\tilde{f}\|, \|\mathrm{D}_x \tilde{f}\| \le \zeta$ small. The higher derivatives $\mathrm{D}_x^j T_Y$, too, contain

[17] It would probably be more natural to consider the higher derivatives as maps into Banach manifolds with exponents $k\rho$, but these norms are equivalent anyways.

a factor $\mathrm{D}_x^i \tilde{f}$ with $0 \le i \le j$ in each term, so by Proposition 3.14 these can be made small. Hence, $\mathrm{D}^{k-1}\Theta^\infty$ and consequently $\|\tilde{h}\|_{k-1}$ can be made uniformly small. Note that $\|\tilde{h}\|_k$ cannot be made small though, see Remark 3.15.

Chapter 4
Extension of Results

In this chapter we discuss some ways to extend the main result of Theorem 3.1 to slightly more general situations. These extensions are known from the compact and Euclidean settings, but a bit scattered over the literature. We try to collect a number of these results here, while extending them to our noncompact setting.

4.1 Non-Autonomous Systems

We proved the main theorem for an autonomous system and perturbation. The Perron method admits without difficulty a time-dependent formulation; we refrained from including this, since it would only have cluttered the already detailed proof, while time-dependence is easily added as an afterthought, as already noted in Sect. 1.6.1.

Let us assume that M is an r-NHIM for the (time-independent) vector field v on (Q, g) and that all assumptions of Theorem 3.1 are fulfilled. We can allow time-dependent perturbations by the standard trick to extend the phase space of the system by $\mathbb{R} \ni t$. Define

$$\hat{Q} = \mathbb{R} \times Q \qquad \text{with metric} \quad \hat{g} = \mathrm{d}t^2 + g. \tag{4.1}$$

Then $(\hat{Q}, \hat{g})$ is again of bounded geometry. We trivially extend the vector field v to

$$\hat{v}(t, x) = \big(1, v(x)\big) \in \mathrm{T}_{(t,x)}\hat{Q} \tag{4.2}$$

and set $\hat{M} = \mathbb{R} \times M \subset \hat{Q}$. Then the flow $\hat{\Phi}$ of $\hat{v}$ has the same hyperbolicity properties as Φ since the additional flow along $\dot{t} = 1$ is completely neutral and decoupled from the original system. It follows that $\hat{M}$ is again an r-NHIM for the dynamical system $(\hat{Q}, \hat{\Phi}, \mathbb{R})$. Note that we need a theory for noncompact NHIMs to perform this extension by the time interval $\mathbb{R}$. Now we can choose a perturbed vector $\tilde{v}$ that depends explicitly on time, as long as $\tilde{v} \in C^{k,\alpha}_{b,u}(\hat{Q})$ is close to $\hat{v}$. This means that the perturbation must be small in $C^{k,\alpha}$-norm (including derivatives with respect to

J. Eldering, *Normally Hyperbolic Invariant Manifolds*, Atlantis Series in Dynamical Systems 2, DOI: 10.2991/978-94-6239-003-4_4,

time), uniformly for all time. As a result we find that the perturbed manifold $\tilde{M}$ will depend on time, i.e. it is not exactly of the form $\tilde{M} = \mathbb{R} \times \mathcal{M}$ for some $\mathcal{M} \subset Q$. We do find that $\tilde{M}$ is uniformly close to $\hat{M} = \mathbb{R} \times M$, however, so $\tilde{M}$ is approximately of this product form.

Remark 4.1 A direct application of Theorem 3.1 requires the perturbed vector field to be $C^{k,\alpha}$ with respect to time, too, since $t \in \mathbb{R}$ is added to the phase space variables. Note that the result thus depends $C^{k,\alpha}$ smoothly on time as well. A closer inspection of the proof shows that this can in fact be replaced by the condition that $\hat{v}(t, \cdot) \in C^{k,\alpha}_{b,u}$, uniformly in $t \in \mathbb{R}$, just as in Remark A.7. In that case the resulting manifold $\tilde{M}$ cannot be expected to be differentiable with respect to time anymore, but it still satisfies all uniform $C^{k,\alpha}$ smoothness and boundedness properties with respect to $x \in Q$. In particular, $\tilde{M}$ is still uniformly close to $\hat{M}$, uniformly for all $t \in \mathbb{R}$. ◇

Instead of starting with an autonomous system v, we can also take an initial non-autonomous system $\hat{v}$ and perturb that. As long as $\hat{v}$ truly describes a non-autonomous system, that is, it is defined on a space $\mathbb{R} \times Q$ and has component 1 along $\mathbb{R}$, then normal hyperbolicity is easily tested. The $\mathbb{R}$-component of the flow is trivially neutral, while the other Q-component must be checked in a context where, for example, also the invariant splitting (1.9) may depend on time, but this introduces no fundamental changes.

4.2 Smooth Parameter Dependence

Another interesting question for applications is if the persistent manifold depends smoothly on the perturbation parameter. This result can be obtained in a similar way as time-dependence, now adding a parameter $p \in P$ to the phase space with trivial dynamics $\dot{p} = 0$. The noncompact theory is not essential here, but it does allow for a simple proof.

Let again (Q, g) and $v = v(p, x)$ describe the system, where $p \in P$ denotes the parameter. For simplicity we assume that $P = \mathbb{R}^n$ and that $p = 0$ corresponds to the unperturbed system for which we have M as r-NHIM. We consider again an extended system $\hat{Q} = P \times Q$ and $\hat{M} = P \times M$. The extended vector field we choose slightly differently: we use an external scaling parameter $\alpha \geq 0$ to slowly 'turn on' the parameter dependence. Let $\chi \in C^\infty(\mathbb{R}_{\geq 0}; [0, 1])$ be a radial cut-off function such that $\chi(r) = 1$ for $r \leq 1$ and $\chi(r) = 0$ for $r \geq 2$, and define

$$\hat{v}_\alpha(p, x) = \big(0, v\big(\chi(\|p\|)\, \alpha\, p, x\big)\big) \tag{4.3}$$

as a vector field on $\hat{Q}$. Note that $\hat{M}$ is an r-NHIM for $\hat{v}_0$ by trivial extension. One can verify that $\|\hat{v}_\alpha - \hat{v}_0\|_r$ can be chosen small with α. Uniformity with respect to p follows automatically from χ having compact support. As a result of Theorem 3.1 we conclude that there exists an $\alpha > 0$ such that $\hat{v}_\alpha$ has a C^r family of invariant manifolds

$$\tilde{M} = \coprod_{p' \in P} \tilde{M}_{p'}, \tag{4.4}$$

where $\tilde{M}_{p'}$ is the invariant manifold corresponding to the vector field $v(p,\ \cdot\,)$ with $p = \chi(\|p\|)\,\alpha\,p'$. This parametrizes a full neighborhood $B(0;\alpha) \subset P$.

4.3 Overflowing Invariant Manifolds

Overflowing invariance is a useful tool to study invariant manifolds whose normal hyperbolicity properties break down beyond a certain domain, see also Sect. 1.6.3. We shall indicate here how our main result can be extended to overflowing invariant manifolds. We provide conditions for persistence that are slightly weaker than those in the literature. These might prove useful for some applications.

The following definition extends that in [Fen72] and is equivalent to Definition 2.1 in [BLZ99].

Definition 4.2 (Overflowing invariant manifold) *Let (Q, g) be a Riemannian manifold, $M \subset Q$ a C^1 submanifold with boundary $\partial M \in C^1$, and $v \in C^1$ a vector field on Q with flow Φ. Let n denote the outward normal at ∂M. Then M is called overflowing invariant under v if the following hold:*

1. *backward orbits stay in M, i.e. $\forall\, m \in M,\ t < 0 : \Phi^t(m) \in M$;*
2. *the vector field v points uniformly strictly outward at ∂M, i.e. there exists some $\varepsilon > 0$ such that $\forall\, m \in \partial M : g_m(v, n) \geq \varepsilon$.*

Definition 1.8 of normal hyperbolicity can be adapted to this setting (only condition (1) is necessary): we assume that only stable normal directions are present and we only require M to be negatively invariant, while the exponential rate conditions must hold along orbits as long as they stay inside M.

Remark 4.3 Note that the uniformity in condition (2) reduces to the standard 'strictly outward' if $(\bar{M}) = M \cup \partial M$ is compact. This is the natural generalization for noncompact manifolds, since the condition is used to guarantee that under small perturbations and in a small tubular neighborhood the vector field is still pointing outward. ◇

The Perron method uses orbits as fundamental objects and constructs a contraction operator on these. The essence of Definition 4.2 is to guarantee condition (1) that backward orbits stay inside $(\bar{M})$, even under a small perturbation of the vector field. This provides an idea to slightly weaken the overflow invariance definition into an a priori argument. If any orbits considered in the Perron method proof stay inside $\tilde{M}$, then all assumptions throughout the proof are still valid and we obtain a persistent manifold $\tilde{M}$. To make this idea explicit, we choose the trivial bundle setting of Theorem 3.2 and introduce the following weakened definition.

Definition 4.4 (A priori overflowing invariance) *Let (X, g) be a Riemannian manifold and let $M \subset X$ an open submanifold, i.e. of the same dimension, with boundary $\partial M \in C^1$. Let Y be a Banach space, and $v \in C^1$ a vector field on $X \times Y$ with flow Φ. Let n denote the outward normal at ∂M. Let $\tilde{v}$ be a perturbation of v. Then M is called a priori overflowing invariant for the pair $(v, \tilde{v})$ if the following hold:*

1. *backward orbits of v stay in M, i.e. $\forall\, m \in M,\ t < 0 : \Phi^t(m) \in M$;*
2. *the vector field $\tilde{v}$ points (non-strictly) outward at a tubular neighborhood over ∂M, i.e. there exists some $\eta > 0$ such that*

$$\forall\,(m, y) \in \partial M \times Y_{\leq \eta} : g\big(\mathrm{D}\pi_X \cdot v(m, y), n(m)\big) \geq 0.$$

Remark 4.5 Note that Definition 4.2 implies 4.4 when $\|\tilde{v} - v\|_1$ is small enough and $\tilde{v} \in C^1_{b,u}$.

Remark 4.6 A useful generalization of Definition 4.4 to the setting of Theorem 3.1 is less trivial. There we do not have canonical vertical fibers over ∂M in the tubular neighborhood, nor the associated projection of v onto $\mathrm{T}X$ at ∂M. We cannot simply take a non-vertical fiber; the Perron method adapts the curves x and y separately, so it may happen that while $x(0) \in \partial M$ is kept fixed, $y(0)$ is updated to a new value such that $(x(0), y(0))$ lies outside of the tubular neighborhood over $\bar{M}$, and control is lost. ◇

Let us demonstrate the application of this more general definition with the following simple example, see also Fig. 4.1.

Example 4.7 (Persistence under a priori overflowing invariance) Let $X \times Y = \mathbb{R} \times \mathbb{R}$ and let the unperturbed vector field be given by

$$v(x, y) = \big(-(x-1)^2,\ (x^2-4)\,y\big).$$

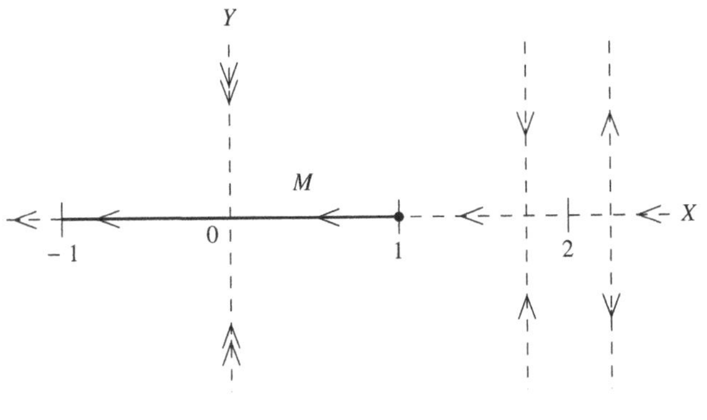

Fig. 4.1 A priori overflowing invariance for the manifold M

Note that $M = (-1, 1) \subset X$ is strictly overflowing invariant at its left boundary $x = -1$ (we could choose other values as well), but non-strictly so at the right boundary $x = 1$, which is a degenerate stationary point. The vector field is normally attracting over the interval $(-2, 2)$ and uniformly so over any closed subinterval. Note that there does not exist a subinterval of X that is overflowing invariant according to Definition 4.2.

Let us choose a family v_δ of perturbations of v such that $\|v_\delta - v\|_1 \leq \delta$ and $v = v_\delta$ on a neighborhood of $(1, 0) \in X \times Y$. Then M satisfies Definition 4.4 for this family v_δ and application of Theorem 4.8 below shows that for δ sufficiently small, there exists a unique negatively invariant manifold $\tilde{M} = \mathrm{Graph}(\tilde{h})$ for the flow of v_δ such that $\tilde{h} : [-1, 1] \subset X \to [-\eta, \eta] \subset Y$. For any $r \geq 1$ there exists a δ such that $\tilde{h} \in C^r$ holds. ○

Theorem 4.8 (Persistence under overflowing invariance) *Let $k \geq 2$, $\alpha \in [0, 1]$ and $r = k + \alpha$. Let (X, g) be a smooth, complete, connected Riemannian manifold of bounded geometry and Y a Banach space. Let $v_\delta \in C^{k,\alpha}_{b,u}$ be a family of vector fields defined on a uniformly sized neighborhood of the zero-section in $X \times Y$ such that $\|v_\delta - v_0\|_1 \leq \delta$. Let M satisfy Definition 4.4 for the pair (v_0, v_δ) for any $\delta \in (0, \delta_0]$ and let M be r-normally attracting for the flow defined by v_0, that is, M satisfies the overflowing invariant version of Definition 1.11 with* $\mathrm{rank}(E^+) = 0$.

Then for each sufficiently small $\eta > 0$ there exist $\delta_1 > 0$ such that for any $\delta \in (0, \delta_1]$, there is a unique manifold with boundary $\tilde{M} = \mathrm{Graph}(\tilde{h})$, $\tilde{h} : M \to Y$, $\|\tilde{h}\|_0 \leq \eta$ such that $\tilde{M}$ is negatively invariant under the flow defined by v_δ. Moreover, $\tilde{h} \in C^{k,\alpha}_{b,u}$ and $\|\tilde{h}\|_{k-1}$ can be made arbitrary small by choosing $\|v_\delta - v_0\|_{k-1}$ sufficiently small. The function h extends continuously to ∂M.

Remark 4.9 In this overflowing invariance setting, the condition that $\mathrm{rank}(E^+) = 0$ is really necessary and not an artifact of our proof. The same results hold for inflowing invariance with no stable normal directions present. Definition 4.4 can be extended to full normal hyperbolicity with both stable and unstable normal directions present. This requires full invariance of a tubular neighborhood of M under both the forward and backward orbits.

Remark 4.10 We can restrict to a smaller open subset U of X that contains $\bar{M}$, so we do not need $v_\delta \in C^{k,\alpha}_{b,u}$ to hold on all of X. If this subset U is not convex, though, we may run into difficulties when applying the mean value theorem, see Fig. 4.2: an intermediate point $\xi \notin \bar{M}$ on the line between x_1, x_2 may be selected, so we need

Fig. 4.2 A nonconvex subset $M \subset X$

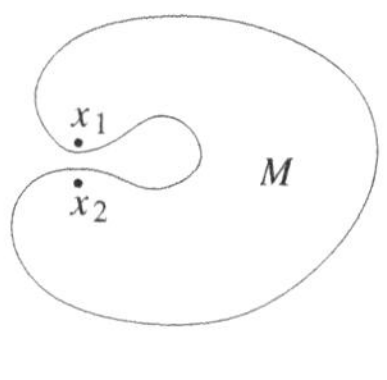

to make sure that the uniform estimates still hold there. Thus the need for $U \supset \bar{M}$ to be convex, see also the remark in [Hen81, p. 289]. ◇

Proof The proof of Theorem 3.2 requires minimal changes. Note that regardless of the modifications and smoothing preparations performed in Sect. 3.4, the vector field $\tilde{v}_X$ is precisely the horizontal component of the perturbed vector field v_δ. In Sect. 3.6 where we proved existence and uniqueness of $\tilde{M}$, we take η small enough that it satisfies condition (2) of Definition 4.4. This guarantees that $x = T_X(y, x_0)$ is a solution curve such that $x\big((-\infty, 0]\big) \subset \bar{M}$ for any $y \in B_\eta^\rho(I; Y)$ and $x_0 \in \bar{M}$. Hence, the contraction mapping $T = T_Y \circ (T_X\,,\ \mathrm{pr}_1)$ is well-defined with intermediate space $\mathcal{B}_\beta(I; \bar{M})$ and we find a unique Lipschitz continuous fixed point map Θ^∞ : $\bar{M} \to B_\eta^\rho(I; Y)$.

No essential changes are needed with respect to the smoothness proof in Sect. 3.7. The formal derivatives (3.43) are well-defined along all curves x and y that are considered, since the derivatives of $\tilde{v}_X$, A, f are defined on an open neighborhood of $\bar{M}$. In Sect. 3.7.7 we use the mean value theorem to prove that the restricted maps $T^{b,a}$ have true derivatives. Remark 4.10 is not problematic here, since $T^{b,a}$ is defined on the finite interval $J = [a, 0]$ and thus we can restrict to arbitrarily small open neighborhoods along the curves x, y when restricted to J. Hence we find that $\Theta^\infty \in C_{b,u}^{k,\alpha}$ on M. □

4.4 Full Normal Hyperbolicity

We made the assumption in our main theorems that the unstable bundle E^+ was absent, that is, that M was a normally attracting invariant manifold. As already noted in Remark 3.3, 8, it should be possible to generalize this to the case of full normal hyperbolicity where both stable and unstable normal directions are present. Let us indicate here how this more general result can be obtained.

Assume that in Theorem 3.1 we have an invariant splitting (1.9) with both stable and unstable bundles $E^\pm$ present. The reduction principle in Sect. 2.6 leads to a formulation of Theorem 3.2 with a trivial bundle

$$\pi : X \times \big(Y \times Z\big) \to X, \tag{4.5}$$

where X is a smoothed approximation of M and approximate, boundedly smooth representations of the bundles $E^\pm$ are embedded into the trivial bundles $X \times Y$, $X \times Z$ with Y, Z Banach spaces. This means that M is again represented as the graph of an approximate zero section $h_\sigma : X \to Y \times Z$; now, the subbundles $X \times Y$ and $X \times Z$ are approximately invariant under v_σ. The deviation from invariance is controlled by σ, the parameter of the smoothing approximation of M. We find linear operators $A^\pm(x)$ on Y and Z respectively, that approximate the linearizations of v_Y and v_Z, and corresponding flows $\Psi^{\pm}$ with approximate growth rates. We add a map[1]

$$T_Z(x, y, z)(t) = \int_t^{\infty} \Psi_x^+(t, \tau)\, \tilde{f}^+\big(x(\tau), y(\tau), z(\tau)\big)\mathrm{d}\tau \tag{4.6}$$

with $z \in B_\eta^\rho(\mathbb{R}_{\geq 0}; Z)$ and adapt the other maps to incorporate z as an argument. We use Lemma 3.30 and extend all curves in $X,\ Y,\ Z$ to the full real line. This should yield a contraction

$$T = (T_Y, T_Z) \circ \big(T_X\,,\ \mathrm{pr}_1\,,\ \mathrm{pr}_2\big) \quad \text{on} \quad B_\eta^\rho(\mathbb{R}; Y) \times B_\eta^\rho(\mathbb{R}; Z), \tag{4.7}$$

again with $x_0 \in X$ as initial value parameter. We obtain a pair $(\Theta^-,\ \Theta^+)$ of fixed point maps, and after evaluation at $t = 0$ we find

$$(\tilde{h}^-, \tilde{h}^+) : X \to Y \times Z, \tag{4.8}$$

whose graph describes the persistent invariant manifold $\tilde{M}$. See e.g. [VG87] for an application of this technique in the center manifold setting.

An alternative method to obtain $\tilde{M}$ is by constructing it as the intersection of its (center-)stable and (center-)unstable manifolds. Here we make use of over- and inflowing invariance. First we construct a tubular neighborhood $P_{\leq\eta}^+$ of M along (a smooth approximation of) the unstable bundle E^+, see Fig. 4.3. This manifold is approximately invariant and overflowing according to Definition 4.2 if η is sufficiently small since it lies approximately along the unstable direction of M. Careful modifications to the setup of the proof in Sects. 3.4 and 3.5 will allow us recover the local unstable manifold W_{loc}^U of $\tilde{M}$ as the graph of

$$\tilde{h}^U : P_{\leq\eta}^+ \to \bar{e}^-,$$

Fig. 4.3 An overflowing tubular neighborhood $P_{\leq\eta}^+$ along the unstable manifold

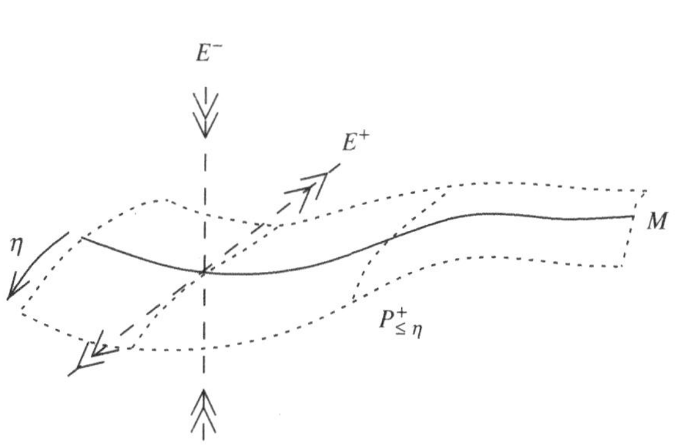

[1] Note that since $t \leq \tau$, we have a reverse flow $\Psi^+(t, \tau)$ for the unstable directions, which indeed satisfies the growth estimates (1.10).

where $\bar{e}^-$ is an extension of the bundle E^- over $P^+_{\leq\eta}$. Vice versa we can find the local stable bundle W^S_{loc} as a graph over $P^-_{\leq\eta}$. Their intersection yields

$$\tilde{M} = W^U_{\text{loc}} \cap W^S_{\text{loc}} \in C^{k,\alpha}_{b,u}$$

since the intersection is uniformly transversal. This is a standard trick which is also applied in [Fen72; HPS77] using the graph transform.

4.5 Recovery of the Invariant Fibration and Splitting

In our main Theorem 3.1 we did not prove that the persistent manifold is again normally hyperbolic. As stated in Remark 3.3, 9, we must recover the invariant splitting of the perturbed manifold $\tilde{M}$ to be able to conclude that $\tilde{M}$ again satisfies Definition 1.8 of normal hyperbolicity. We will sketch how to find the invariant fibers of the stable manifold of $\tilde{M}$ in the setting of Theorem 3.2 with a trivial bundle $Q = X \times Y$. These fibers are easily recovered using the Perron method, and smoothness of each single fiber essentially comes for free. We then readily obtain the vertical part $\tilde{E}^-$ of the invariant splitting as the tangent planes to these fibers at $\tilde{M}$. Since $\mathrm{T}\tilde{M}$ is already invariant, we have

$$\mathrm{T}_{\tilde{M}} Q = \mathrm{T}\tilde{M} \oplus \tilde{E}^- \tag{4.9}$$

which is invariant under the tangent flow, cf. (1.9). Note that we should still prove boundedness and uniform continuity of this splitting.

Each single invariant fiber can be found by application of the non-autonomous Perron method. Let $\tilde{\Upsilon} : \mathbb{R} \times \tilde{M} \to \tilde{M}$ denote the flow of the perturbed system, restricted to the persistent manifold, and let $\gamma(t) = \tilde{\Upsilon}^t(m)$ denote a solution curve in $\tilde{M}$. Consider the pullback bundle

$$\mathcal{E}_m = \gamma^*(\mathrm{T}Q) \xrightarrow{\pi} \mathbb{R}. \tag{4.10}$$

Note that $\gamma^*(\mathrm{T}\tilde{M})$ is a subbundle of $\mathcal{E}_m$ in a natural way, and that it is invariant under the tangent flow $\mathrm{D}\tilde{\Upsilon}^t$. Next we consider $\gamma^*(E^-)$ as a subbundle of $\mathcal{E}_m$: even though E^- was defined over the original manifold M, it can simply be translated along the canonical fiber Y to be identified with a bundle over $\tilde{M}$ and then be pulled back along γ. Thus we have a splitting

$$T_{\tilde{M}} Q = \mathrm{T}_{\tilde{M}}(X \times Y) = \mathrm{T}\tilde{M} \oplus E^-.$$

The original splitting $\mathrm{T}M \oplus E^-$ was bounded, hence when $\tilde{M}$ is sufficiently close to M in C^1 norm, then this splitting will be bounded again. It is uniformly continuous since both subbundles are. In order to recover a smooth stable fibration later, we will

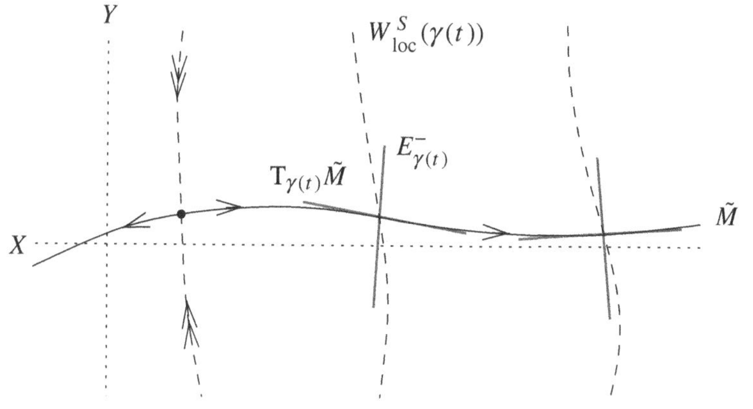

Fig. 4.4 The invariant fibers of $W^S_{\text{loc}}(\tilde{M})$ modeled on the bundle $\mathcal{E}_m = \gamma^*(\mathrm{T}\tilde{M} \oplus \hat{E}^-)$

moreover want to construct a smooth approximation $\hat{E}^-$, similar to the convolution smoothing of $\hat{A}$ in Sect. 3.4. The stable manifold $W^S_{\text{loc}}(\tilde{M})$ of the perturbed system consists of single fibers $W^S_{\text{loc}}(m)$ that can be modeled as graphs of maps

$$\tilde{h}^S_m : \hat{E}^-_m \to \mathrm{T}_m\tilde{M},$$

see Fig. 4.4. We recover the map $\tilde{h}^S_m$ after application of the Perron method to the non-autonomous system defined on $\mathcal{E}_m$ by pulling back the vector field $(\frac{\partial}{\partial t}, \tilde{v})$ along[2]

$$\varphi : \gamma^*(\mathrm{T}Q) \to \mathbb{R} \times Q : (t, x) \mapsto \exp_{\gamma(t)}(x) \quad \text{with} \quad x \in \mathrm{T}_{\gamma(t)}Q \tag{4.11}$$

That is, we look at the system in normal coordinates that follow the solution curve γ. This means that the vector field $\varphi^*(\frac{\partial}{\partial t}, \tilde{v})$ on $\mathcal{E}_m$ has constant speed 1 on the base $\mathbb{R}$, i.e. the time axis, and preserves the zero section, i.e. the origin of the coordinates at $\gamma(t)$. Since $\mathrm{D}\exp_{\gamma(t)}(0) = \mathbb{1}$, the exponential growth conditions are preserved under pullback. Hence, the splitting

$$\mathcal{E}_m = \gamma^*(\mathrm{T}\tilde{M}) \oplus \gamma^*(\hat{E}^-)$$

is approximately invariant with growth rates close to ρ_X and ρ_Y respectively. Now we apply the Perron method on the interval $\mathbb{R}_{\geq 0}$. That is, we have the operator

[2] We only can (and need to) perform the pullback in an η-sized tubular neighborhood of γ. Outside this neighborhood we smoothly cut off the vector field to a suitable linearization. Therefore we only recover the local stable manifold.

$$T(x, y, y_0)(t) = \left(\int_t^\infty \mathrm{D}\tilde{\Upsilon}^{t-\tau}(\gamma(\tau))\, f_X(\tau, x(\tau), y(\tau))\mathrm{d}\tau, \right. \tag{4.12}$$
$$\left. \Psi_\gamma(t, 0)\, y_0 + \int_0^t \Psi_\gamma(t, \tau)\, f_Y(\tau, x(\tau), y(\tau))\mathrm{d}\tau \right)$$

which is a contraction on pairs of curves $(x, y) \in B^\rho(\mathbb{R}_{\geq 0}; \mathcal{E}_m)$ with the partial initial condition $y_0 \in \mathcal{E}_{m,0} = \hat{E}_m^-$ as parameter. Here we chose $\rho_X < \rho < \rho_X$ and used the fact that $f_{X,Y}(t, 0, 0) = 0$. The functions $f_{X,Y}$ are C^1-small nonlinear terms. These contain the linear and higher nonlinear perturbations away from the linear flows $\mathrm{D}\Upsilon$ and Ψ_γ along $\gamma^*(\mathrm{T}\tilde{M})$ and $\gamma^*(\hat{E}^-)$ respectively, corrected for the bounded coordinate changes introduced by φ in (4.11). Thus, we find $\tilde{h}_m^S$ by evaluating the x-component of the fixed point map at $t = 0$ for a given y_0.

We can establish that $\tilde{h}_m^S \in C_{b,u}^{k,\alpha}$ by an implicit function argument[3] on the Banach space $B^\rho(\mathbb{R}_{\geq 0}; \mathcal{E}_m)$, see [Irw70]. The unpublished preprint [Dui76] extends these ideas to the geometric context of Banach bundles over $\mathbb{R}$.

Note that this result is uniform in the family parameter $m \in \tilde{M}$, but it does not include smoothness with respect to m. To obtain smoothness, we extend the pullback bundle (4.10) to

$$\mathcal{E} = \coprod_{m \in \tilde{M}} \mathcal{E}_m = \Upsilon^*(\mathrm{T}Q) \xrightarrow{\pi} \mathbb{R} \times \tilde{M}. \tag{4.13}$$

This bundle can be endowed with a topology by parallel transport along the curves γ as in Sect. 3.7.4. Then each derivative with respect to m adds an exponential growth factor $e^{\rho_{\tilde{M}}\, t}$ since the curves γ can diverge at that rate in forward time. Note that $\rho_{\tilde{M}}$ corresponds to ρ_M in (1.11), that is, the upper bound on the spectrum of the tangential flow. This leads to the spectral gap condition

$$\rho_Y < \rho_X - r\, \rho_{\tilde{M}} \approx (r + 1)\rho_X, \tag{4.14}$$

from which it follows that the mapping $m \mapsto \tilde{h}_m^S$ has $C_{b,u}^{k-1,\alpha}$ smoothness. This implies that the invariant splitting is at least uniformly continuous and bounded. The exponential growth rates with respect to the splitting follow from small perturbation estimates similar to those in Sects. 3.4 and 3.5. This completes the requirements for normal hyperbolicity of $\tilde{M}$ in Definition 1.8.

[3] We immediately recover higher smoothness instead of having to go through an elaborate scheme involving the fiber contraction theorem as in Sect. 3.7. The reason is that $B^{k\rho}(\mathbb{R}_{\geq 0}; \mathcal{E}_m) \hookrightarrow B^\rho(\mathbb{R}_{\geq 0}; \mathcal{E}_m)$ is a continuous embedding when $\rho < 0$ and we look at $\mathbb{R}_{\geq 0}$, that is, higher powers of negative exponential growth decay even stronger.

Appendix A
Explicit Estimates in the Implicit Function Theorem

In this appendix, we carefully examine the implicit function theorem. We extend this standard theorem to classes of functions with additional properties such as boundedness and uniform and Hölder continuity. The crucial ingredient is the explicit formula (A.2) for the derivative of the implicit function, which allows us to transfer regularity conditions onto the implicit function.

As an application of the implicit function theorem in Banach spaces, we will establish existence, uniqueness and smooth dependence on parameters for the flow of a system of ordinary differential equations. Essentially, these are standard results from differential calculus, see e.g. Zeidler [Zei86, p. 150,165] or [Irw72, Rob68]. We consider a general setting of ODEs in Banach spaces and show smooth dependence, both on the initial data, as well as on the vector field itself. Moreover, our extension of the implicit function theorem yields boundedness and uniform continuity results.

We start with some results on inversion of linear maps.

Lemma A.1 (Invertibility of linear maps) *Let X be a Banach space and let $A \in \mathcal{L}(X)$ be a continuous linear operator with continuous inverse. Let $B \in \mathcal{L}(X)$ be another linear operator such that $\|B\| < \frac{1}{\|A^{-1}\|}$. Then $A + B$ is also a continuous linear operator with continuous inverse, given by the absolutely convergent series*

$$(A+B)^{-1} = \sum_{n \geq 0} (-A^{-1}B)^n \, A^{-1} = \sum_{n \geq 0} A^{-1} (-B\,A^{-1})^n. \qquad \text{(A.1)}$$

Proof First of all, note that there exists an $M \geq 1$ such that $\|A\|, \|A^{-1}\| \leq M$. The base of the geometric series can be estimated in operator norm as $\|-A^{-1}B\| < 1$, so the series is absolutely convergent and the limit is a well-defined continuous linear operator, whose operator norm can be estimated as

$$\|(A+B)^{-1}\| \leq \|A^{-1}\| \sum_{n \geq 0} \|-A^{-1}B\|^n \leq \frac{\|A^{-1}\|}{1 - \|A^{-1}B\|} < \infty.$$

J. Eldering, *Normally Hyperbolic Invariant Manifolds*, Atlantis Series in Dynamical Systems 2, DOI: 10.2991/978-94-6239-003-4,

That the limit is again a well-defined linear operator follow from the fact that $\mathcal{L}(X)$ is a Banach space.

Applying $A + B$ to the left-hand side of (A.1), we see that the candidate is a right inverse:

$$(A+B)\sum_{n\geq 0} A^{-1}\,(-BA^{-1})^n = \sum_{n\geq 0}(-BA^{-1})^n - \sum_{n\geq 0}(-BA^{-1})^{n+1} = 1.$$

Similarly the candidate can be shown to be a left inverse of $A + B$. Now we have that the candidate is continuous and a full inverse and furthermore, $A + B$ itself is clearly a continuous operator as the sum of two continuous operators, so the proof is completed. □

Corollary A.2 (Linear inversion is analytic) *Let $I\colon A \mapsto A^{-1}$ be the inversion map defined on continuous, linear mappings $A \in \mathcal{L}(X)$ with continuous inverse, where X is a Banach space. The map I is analytic with radius of convergence $\rho(A) \geq 1/\|A^{-1}\|$. When X is finite-dimensional, I is a fortiori a rational map.*

Proof Extending well-known results on analytic functions to Banach spaces (see e.g. [Muj86]), we read off from (A.1) that the inversion map I can be given around A by an absolutely convergent power series with $\rho(A) \geq 1/\|A^{-1}\|$ and is thus analytic. When X is finite-dimensional, $\det(A) \neq 0$ implies that A^{-1} is a rational expression in the matrix coefficients of A according to Cramer's rule. □

The inversion map I is locally Lipschitz, like every C^1 mapping:

$$\left\|(A+B)^{-1} - A^{-1}\right\| \leq \sum_{n\geq 1}\left\|-A^{-1}B\right\|^n \left\|A^{-1}\right\| \leq \frac{\|A^{-1}\|^2}{1-\left\|A^{-1}B\right\|}\,\|B\|.$$

However, when we restrict to a domain bounded away from non-invertible operators A, that is, when $\|A^{-1}\| \leq M$, then the Lipschitz constant is bounded for small B. This implies that when $A = A(x)$ depends on a parameter via a certain continuity modulus, then $A(x)^{-1}$ will have the same continuity modulus up to the Lipschitz constant, at least in small enough neighborhoods.

The standard implicit function theorem on Banach spaces can be stated as

Theorem A.3 (Implicit function theorem) *Let X be a Banach space, Y a normed linear space, and let $f \in C^{k\geq 1}(X \times Y; X)$. Let $(x_0, y_0) \in X \times Y$ and assume that $f(x_0, y_0) = 0$ and that $\mathrm{D}_1 f(x_0, y_0)^{-1} \in \mathcal{L}(X)$ exists as a continuous, linear operator.*

Then there exist neighborhoods $U \subset X$ of x_0 and $V \subset Y$ of y_0, and a unique function $g\colon V \to U$ such that $f(g(y), y) = 0$. Furthermore, the map g is C^k and the derivative of g is given by the formula

$$\mathrm{D}g(y) = -\mathrm{D}_1 f(g(y), y)^{-1} \cdot \mathrm{D}_2 f(g(y), y). \tag{A.2}$$

See [Zei86, pp. 150–155] for a proof. Note that we do not need to assume that Y is a complete space, as the contraction theorem is only applied on X. Recall that we use notation where D denotes a total derivative, while D_i with index $i \in \mathbb{N}$ denotes a partial derivative with respect to the i-th argument.

Formula (A.2) for the derivative of the implicit function g will be crucial for the extension of the implicit function theorem to many classes of regularity, extending C^k smoothness. We use the Lipschitz estimate for the inversion map and require that the regularity conditions are preserved under composition, addition, multiplication and localization of functions. By Proposition C.3, the derivatives of g are expressed in terms of $\mathrm{D}_1 f(g(y), y)^{-1}$ acting on a polynomial expression of same or lower order derivatives of f and strictly lower order derivatives of g.

As an example, let us take $C_b^{k,\alpha}$ functions. Using Lemma 1.19 and induction over k, this function class is preserved under products. For composition, we check Hölder continuity,

$$\|f(g(x_2)) - f(g(x_1))\| \le C_f \left(C_g \|x_2 - x_1\|^\alpha\right)^\alpha \le (C_f\, C_g^\alpha)\, \|x_2 - x_1\|^\alpha$$

for $0 < \alpha \le 1$, when $\|x_2 - x_1\| \le 1$. In case $\|x_2 - x_1\| > 1$ however, we can directly use the boundedness of f:

$$\|f(g(x_2)) - f(g(x_1))\| \le \|f(g(x_2))\| + \|f(g(x_1))\| \le 2\, \|f\|_0\, \|x_2 - x_1\|^\alpha.$$

Thus, Hölder continuity is preserved with some new Hölder constant, while boundedness is trivially preserved as well. We conclude that if $f \in C_b^{k,\alpha}$, and $(\mathrm{D}_1 f)^{-1}$ is globally bounded, then we can read off from formula (A.2) that $g \in C_b^{k,\alpha}$. The same results hold for the class of $C_{b,u}^k$ functions, or any other class of functions whose properties are preserved when inserted into (A.2). Together, interpreting $\alpha = 0$ as an empty condition, these lead to

Corollary A.4 *Let in the Implicit Function Theorem A.3, $f \in C_{b,u}^{k,\alpha}$ with $k \ge 1$ and $0 \le \alpha \le 1$. Assume moreover that $\left\|\mathrm{D}_1 f(x, y)^{-1}\right\| \le M$ is bounded on $U \times V$ for some constant $M < \infty$. Then $g \in C_{b,u}^{k,\alpha}$, and the boundedness and continuity estimates depend in an explicit way on those of f.*

Remark A.5 Formula (A.2) only provides control on the derivatives of the implicit function, but the size of g itself can be controlled by choice of the neighborhood U. In our applications, this will match up with choosing coordinate charts around the origin in $\mathbb{R}^n$. ◇

Let us now consider an ordinary differential equation

$$\dot{x} = f(t, x), \qquad x(t_0) = x_0, \tag{A.3}$$

where x takes values in a Banach space B and $f \in C_{b,u}^{k,\alpha}(\mathbb{R} \times B; B)$ with $k \ge 1$, $0 \le \alpha \le 1$. We consider solutions $x \in X = C^0(I; B)$ equipped with the supremum norm, which turns X into a Banach space.[1] We choose I to be a closed interval $I = [a, b] \subset \mathbb{R}$. The Picard integral operator

$$T \colon X \to X \colon x(t) \mapsto F(x)(t) = x_0 + \int_{t_0}^{t} f(\tau, x(\tau))\, \mathrm{d}\tau \tag{A.4}$$

has exactly the solution curves of (A.3) as fixed points. It also implicitly depends on $f \in C^{k,\alpha}_{b,u}(\mathbb{R} \times B;\, B)$ and $(t_0, x_0) \in I \times B$. From now on we denote by D_x a partial derivative with respect to the argument that is typically described by the variable x.

This T is a contraction for $|I| = b - a$ small enough:

$$\begin{aligned}
\|T(x_1) - T(x_2)\| &= \sup_{t\in I} \left\| \int_{t_0}^{t} f(\tau, x_1(\tau)) - f(\tau, x_2(\tau))\; \mathrm{d}\tau \right\| \\
&\le \sup_{t\in I} \int_{t_0}^{t} \|\mathrm{D}_x f(\tau, \xi(\tau))\| \|x_1(\tau) - x_2(\tau)\|\, \mathrm{d}\tau \\
&\le \sup_{t\in I} |t - t_0| \|\mathrm{D}_x f\| \|x_1 - x_2\| \\
&\le |I|\, \|\mathrm{D}_x f\|\, \|x_1 - x_2\|.
\end{aligned}$$

We restrict T to a bounded subset of argument functions f,

$$\mathcal{F} \subset C^{k,\alpha}_{b,u}(\mathbb{R} \times B;\, B), \qquad \sup_{f\in\mathcal{F}} \|f\|_{k,\alpha} \le R.$$

Thus, choosing $|I| \le \frac{1}{2R}$ turns T into a $q = \frac{1}{2}$ contraction, which shows that there is a unique $x \in X$ satisfying $T(x) = x$ and therefore (A.3).

Next, we consider small perturbations of both (t_0, x_0) and f. To apply the implicit function theorem, we define $F(x) = x - T(x)$. This function has a unique zero and $\mathrm{D}F(x)$ is invertible, as

$$\begin{aligned}
&F(x + \delta x)(t) - F(x)(t) \\
&= \delta x(t) - \int_{t_0}^{t} \mathrm{D}_x f\big(\tau, \xi(\tau)\big) \cdot \delta x(\tau)\, \mathrm{d}\tau \\
&= \delta x(t) - \int_{t_0}^{t} \mathrm{D}_x f\big(\tau, x(\tau)\big) \cdot \delta x(\tau) + O\big(\|\xi(\tau) - x(\tau)\|\big) \|\delta x(\tau)\|\, \mathrm{d}\tau \\
&= \big(\mathrm{D}F(x) \cdot \delta x\big)(t) + o\big(\|\delta x\|\big)
\end{aligned} \tag{A.5}$$

The neglected terms are $o\big(\|\delta x\|\big)$ since $\mathrm{D}_x f$ is uniformly continuous on I, so $\mathrm{D}F(x)$ exists. From the expression above, we can also easily read off continuity of $\mathrm{D}F(x)$ as a linear operator, by writing $\mathrm{D}F(x) = \mathbb{1} + A(x)$ and noticing that

---

[1] Note that any actual solution x will be C^1 at least, but only $x \in C^0$ is required. This makes X a complete space without the need to introduce norms more complicated than the supremum norm.

$$\|A(x)\| \leq |I| \, \|\mathrm{D}_x f\| < \tfrac{1}{2},$$

thus $\mathrm{D}F(x)$ is a bounded, invertible linear operator such that $\|\mathrm{D}F(x)^{-1}\| \leq 2$.

By similar estimates, the derivatives of F with respect to the parameters t_0, x_0, and f can be calculated as

$$\begin{aligned} \mathrm{D}_{t_0} F(x) &= f(t_0, x(t_0)), \\ \mathrm{D}_{x_0} F(x) &= -\mathbb{1}, \\ \big(\mathrm{D}_f F(x) \cdot \delta f\big)(t) &= -\int_{t_0}^{t} \delta f(\tau, x(\tau)) \, \mathrm{d}\tau. \end{aligned} \tag{A.6}$$

Note that these are all bounded linear operators; $\mathrm{D}_{t_0} F(x)$ is because $\|f\| \leq R$. Hence, $F \in C_b^1$ as a function of x, t_0, x_0, f, so by the implicit function theorem, the solution $x(t; t_0, x_0, f)$ depends C_b^1 on t_0, x_0, f.

Next, we establish $C_{b,u}^{k,\alpha}$ dependence on the initial conditions t_0, x_0 and C_b^k dependence on f and t_0, x_0 together. Uniform and Hölder dependence on f are lost because the variations $\delta f \in C_{b,u}^{k,\alpha}$ are not uniformly equicontinuous. The first derivatives can be differentiated another $k-1$ times with respect to each of the variables, using similar estimates as in (A.5). These derivatives are continuous as f is uniformly continuous on the interval I. Uniform and Hölder continuity with respect to t_0, x_0 can be read off directly from the expressions (A.5), (A.6) or their higher order derivatives, as $f \in C_{b,u}^{k,\alpha}$. The implicit function theorem only gives an explicit formula (A.2) for the derivative. Here, this translates into the fact that no boundedness follows for the C^0-norm of the solution curve, only for the norms on the derivatives.

We have thus shown that the conditions of Corollary A.4 of the implicit function theorem have been satisfied, so there exists a neighborhood of (t_0, x_0, f) in $I \times B \times \mathcal{F}$ such that for each (t_0', x_0', f') in that neighborhood there is a unique solution to (A.3) and the solutions x depend in a $C_{b,u}^{k,\alpha}$ way on t_0', x_0' and C_b^k on all of t_0', x_0', f'. Note that this result is obtained only on the interval I. We can however extend these results to any bounded interval, by using the composition property of a flow; the estimates may grow with interval size though. Hence, we have the following result, see also [DK00, App. B].

Theorem A.6 (Uniform dependence on parameters of ODE solutions) *Let an ordinary differential equation (A.3) be given, where $f \in \mathcal{F} \subset C_{b,u}^{k,\alpha}(\mathbb{R} \times B; B)$ with $k \geq 1$, $0 \leq \alpha \leq 1$, B a Banach space, and $\mathcal{F}$ a bounded subset. Let $I \subset \mathbb{R}$ be a bounded interval and $X = C^0(I; B)$ the Banach space of (solution) curves, endowed with the supremum norm.*

Then the flow Φ is a C_b^k mapping

$$\Phi \colon I \times B \times \mathcal{F} \to X \colon (t_0, x_0, f) \mapsto \big(t \mapsto x(t)\big).$$

The boundedness is understood to hold only for the derivatives. Moreover, $\Phi \in C_{b,u}^{k,\alpha}$ holds as a mapping from $I \times B$ for fixed $f \in \mathcal{F}$.

Remark A.7 Differentiable dependence on time can be dropped from this theorem. That is, let us instead assume that $f(t, x)$ and its derivatives $\mathrm{D}_x^i f(t, x)$, $i \leq k$ with respect to x are bounded continuous with respect to (t, x). Then the flow is a C_b^k mapping

$$\Phi\colon B \times \mathcal{F} \to X\colon (x_0, f) \mapsto \big(t \mapsto x(t)\big)$$

when $I \subset \mathbb{R}$ is a bounded interval. This result follows directly from the proof, since we only used differentiability with respect to t for differentiable dependence of Φ on t.

Remark A.8 Instead of a Banach space B, we can also choose the setting of a Riemannian manifold (M, g). Solving for the flow of a differential equation is defined in terms of local charts, so by standard arguments the C^k smoothness result extends to this setting.

If we assume moreover in the context of Chap. 2 that (M, g) has bounded geometry and that $f \in C_{b,u}^{k,\alpha}$, then we can obtain stronger results close to those of Theorem A.6. In any single normal coordinate chart the results of Theorem A.6 hold. To extend the flow beyond one chart, we use the fact that coordinate chart transitions are uniformly C^k-bounded maps. It follows that $\Phi \in C_{b,u}^{k,\alpha}$ on any domain such that all image curves are covered by a uniformly bounded number of charts. This includes the domain $M \times I$ for any finite interval $I \subset R$, since f itself is assumed bounded. The bounds and continuity moduli will depend on $|I|$ though.

Alternatively, uniform (Hölder) continuity estimates independent of charts can be obtained by using Proposition 2.13 to express continuity moduli in terms of parallel transport. See Lemma C.10, which is proven via a variation of constants method. ◇

Appendix B
The Nemytskii Operator

The Nemytskii operator creates a mapping on curves from a simple function between spaces. That is, in its simplest form, if we have a function $f : \mathbb{R}^n \to \mathbb{R}^m$, then the associated Nemytskii operator

$$F : C(\mathbb{R}; \mathbb{R}^n) \to C(\mathbb{R}; \mathbb{R}^m), \qquad F(x)(t) = f(x(t)),$$

maps curves x in $\mathbb{R}^n$ to curves $y = F(x) = f \circ x$ in $\mathbb{R}^m$. See also [Van89, pp. 103–109] for a clear presentation.

We investigate continuity of the Nemytskii operator for certain classes of curves. The following definition of the Nemytskii operator in a somewhat more abstract context on bundles over $\mathbb{R}$ allows e.g. for the map f to be time-dependent.

Definition B.1 (Nemytskii operator) *Let $I \subset \mathbb{R}$ and let X, Y be normed vector bundles*[2] *over I. Furthermore, let $f : X \to Y$ be a bundle map, i.e. a fiberwise mapping that covers the identity on I, but which is not necessarily linear in the fibers. We define the corresponding Nemytskii operator*

$$F : \Gamma(X) \to \Gamma(Y) : x \mapsto f \circ x, \tag{B.1}$$

mapping continuous sections of X to continuous sections of Y.

In the previous definition as well as in the following lemma, we need not restrict to vector bundles; we shall also require the case that X is a trivial fiber bundle with a metric space as fiber (e.g. the bundle $\mathcal{B}^\rho_\beta(I; X)$ in the context of Chap. 3). Recall that the space of sections $\Gamma(X)$ can be endowed with an exponential growth distance (1.17) or norm (1.16), respectively. This turns $\Gamma(X)$ into a metric (or normed linear) space denoted by $\Gamma^\rho(X)$ with exponent $\rho \in \mathbb{R}$. The distance $d_\rho(x_1, x_2)$ may

---

[2] For our purposes, a sufficient definition of a normed vector bundle $\pi : X \to \mathbb{R}$ is that there exist local trivializations $\tau : \pi^{-1}(U) \to U \times F$ that are isometric with respect to the norms on X and the normed linear space F. Note that we canonically have such trivializations by parallel transport, see (3.52) and Proposition 3.34.

J. Eldering, *Normally Hyperbolic Invariant Manifolds*, Atlantis Series in Dynamical Systems 2, DOI: 10.2991/978-94-6239-003-4,

be infinite for some x_1, $x_2 \in \Gamma^{\rho}(X)$ if X is a trivial metric fiber bundle. This is not a problem, since it is only used to obtain (local) continuity estimates for sections such that $d_{\rho}(x_1, x_2) < \infty$.

Lemma B.2 (Continuity of the Nemytskii operator) *Let X, Y be normed vector bundles over $I = \mathbb{R}_{\geq 0}$, or alternatively let X be a trivial fiber bundle of a metric space. Let $f \in C^0(X; Y)$ be a continuous fiberwise mapping and let $F : \Gamma(X) \to \Gamma(Y)$ be defined as in (B.1). Let $\rho_1, \rho_2 \in \mathbb{R}$ and assume that one of the following holds:*

1. *$\rho_2 > 0$ and f is bounded into the normed vector bundle Y;*
2. *$\rho_2 \geq \alpha\, \rho_1$ and f is α-Hölder continuous with $0 < \alpha \leq 1$, uniformly with respect to the fibers.*

Then F is continuous as a map $\Gamma^{\rho_1}(X) \to \Gamma^{\rho_2}(Y)$ and under (2), F is moreover α-Hölder continuous again.

Proof We first prove the statement under assumption (i). Fix $x_1 \in \Gamma^{\rho_1}(X)$, let $\varepsilon > 0$ be given, and let $x_2 \in \Gamma^{\rho_1}(X)$ be arbitrary. As f is bounded and $\rho_2 > 0$, we can choose a $T > 0$ such that

$$\forall\, t > T\colon \|f(x_1(t)) - f(x_2(t))\|\, e^{-\rho_2 t} \leq 2\, \|f\|\, e^{-\rho_2 T} \leq \varepsilon.$$

This leaves only the compact interval $[0, T]$ for which we still have to show that $\|f(x_1(t)) - f(x_2(t))\|\, e^{-\rho_2 t} \leq \varepsilon$. Let us denote $g\colon I \to \mathbb{R}\colon t \mapsto e^{-\rho_2 t}$, then the continuity estimate of $f \cdot g\colon X \to Y$ is uniform on the compact set $x_1([0, T])$. Hence, there exists a $\delta' > 0$ such that for all $t \in [0, T]$ and $\xi_2 \in \pi^{-1}(t) \subset X$,

$$d(x_1(t), \xi_2) \leq \delta' \implies e^{-\rho_2 t}\, \|f(x_1(t)) - f(\xi_2)\| \leq \varepsilon.$$

We have that $d(x_1(t), x_2(t)) \leq e^{|\rho_1 T|}\, d_{\rho_1}(x_1, x_2)$, so choosing $\delta = e^{-|\rho_1 T|}\, \delta'$ yields the required estimate for $d_{\rho_1}(x_1, x_2) \leq \delta$. This proves that F is continuous at x_1.

Secondly, assume (2) and let C_α be the Hölder coefficient of f. Then we can estimate

$$\begin{aligned}\|F(x_1) - F(x_2)\|_{\rho_2} &= \sup_{t \geq 0}\, e^{-\rho_2 t}\, \|f(x_1(t)) - f(x_2(t))\| \\ &\leq \sup_{t \geq 0}\, e^{-\rho_2 t}\, C_\alpha \left(d_{\rho_1}(x_1, x_2)\, e^{\rho_1 t}\right)^{\alpha} = C_\alpha\, d_{\rho_1}(x_1, x_2)^{\alpha},\end{aligned}$$

which shows that F is α-Hölder continuous again with coefficient C_α. □

Corollary B.3 *Let the assumptions of Lemma B.2 with condition (1) be satisfied. If f is fiberwise uniformly continuous with continuity modulus independent of the fiber, then also F is uniformly continuous.*

Proof This follows easily: in the proof above, the uniform continuity on the compact set $x_1([0, T])$ can be replaced by the uniform continuity modulus of f itself. This does not depend on x_1, x_2 anymore, only on their distance, so it leads to a uniform continuity modulus of F. □

Remark B.4 The previous results also hold under time inversion. That is, if we consider the interval $I = \mathbb{R}_{\leq 0}$ and invert the inequalities for $\rho_1,\ \rho_2$ in conditions (1) and (2), then Lemma B.2 and Corollary B.3 still hold true. We use this time inverted version in Chap. 3. ◇

Appendix C
Exponential Growth Estimates

In this appendix we investigate the growth rate of higher order derivatives of a general flow on a Riemannian manifold. Basically, if the growth of the tangent flow is proportional to $\exp(\rho\, t)$, then the growth of the r-thorder derivative is of order $\exp(r\, \rho\, t)$. This even extends to 'fractional' derivatives, that is, the $C^{k,\alpha}$-norm (which includes α-Hölder continuity bounds) has this growth behavior for $r = k + \alpha$. These results will be used to obtain continuity and higher order smoothness of the persisting NHIM. The particular exponential growth behavior $\exp(r\, \rho\, t)$ will precisely prescribe the spectral gap condition: to construct a contraction on the r-thderivative, the normal contraction of order $\exp(\rho_Y\, t)$ must dominate the higher order $\exp(r\, \rho_X\, t)$ along the invariant manifold, hence $\rho_Y < r\, \rho_X$ is required.[3]

These results are based on estimating variation of constants integrals and similar in spirit to Gronwall's lemma. We work on Riemannian manifolds, however. This complicates matters with a lot of technicalities, but the basic ideas are still the same. We do require uniform bounds and bounded geometry of the manifold, see Chap. 2. Let us first show the idea for a flow on $\mathbb{R}^n$ and then introduce some concepts and notation to finally treat the general case.

Lemma C.1 (Exponential growth estimates for a flow) *Let $\Phi^{t,t_0} \in C^{k\geq 1}$ be the flow of a time-dependent vector field v on $\mathbb{R}^m$. Let $v(t, \cdot) \in C^k_b(\mathbb{R}^m)$ with all derivatives jointly continuous in $(t, x) \in \mathbb{R} \times \mathbb{R}^m$ and uniformly bounded by $V < \infty$. Suppose that $\|\mathrm{D}\Phi^{t,t_0}(x)\| \leq C_1\, e^{\rho(t-t_0)}$ for all $x \in \mathbb{R}^m$, $t \geq t_0$ and fixed $C_1 > 0$, $\rho \neq 0$. Then for each n, $1 \leq n \leq k$ there exists a bound $C_n > 0$ such that*

$$\forall x \in \mathbb{R}^m,\ t \geq t_0 \colon\ \|\mathrm{D}^n\Phi^{t,t_0}(x)\| \leq \begin{cases} C_n\, e^{n\,\rho(t-t_0)} & \text{if } \rho > 0, \\ C_n\, e^{\rho(t-t_0)} & \text{if } \rho < 0. \end{cases} \tag{C.1}$$

[3] We formulate all statements in this section with respect to exponentially bounded flows in the (more natural) forward time direction. That is, we work with $t \in \mathbb{R}_{\geq t_0}$ and typical exponents $\rho > 0$. In our applications in Chap. 3 we use the time-reversed statements. See also Remark 1.18.

J. Eldering, *Normally Hyperbolic Invariant Manifolds*, Atlantis Series in Dynamical Systems 2, DOI: 10.2991/978-94-6239-003-4,

Proof Let D denote the partial derivative with respect to the spatial variable $x \in \mathbb{R}^m$. We suppress the time dependence in the notation of v since we have the bound $\|D^n v\| \leq V$ for all $1 \leq n \leq k$, uniformly in space and time.

Since Φ^{t,t_0} is a flow, we have

$$D\Phi^{t_0,t_0}(x) = \mathbb{1} \quad \text{and} \quad D^n \Phi^{t_0,t_0}(x) = 0, \quad 2 \leq n \leq k. \tag{C.2}$$

For $1 \leq n \leq k$ we can write, suppressing arguments t_0, x,

$$\frac{d}{dt} D^n \Phi^t = D^n(v \circ \Phi^t) = Dv \circ \Phi^t \cdot D^n \Phi^t + \sum_{l=2}^{n} D^l v \circ \Phi^t \cdot P_{l,n}\big(D^1 \Phi^t, \ldots, D^{n-1} \Phi^t\big), \tag{C.3}$$

where the $P_{l,n}$ are homogeneous, weighted polynomials as in Definition C.2 below. In the first equality, the switching of partial derivatives is well-defined, because the spatial derivative in the middle expression is well-defined and the resulting function continuous. In the right-hand expression we have already used Proposition C.3 and separated the homogeneous term with $D^n \Phi^t$ (when $l = 1$). The result is a linear differential equation for $D^n \Phi^t$ with the inhomogeneous terms in the sum consisting of lower order derivatives $D^i \Phi^t$, $i < n$, only.

For $n = 1$, statement (C.1) is already true by assumption and in that case we also see that (C.3) is a homogeneous linear differential equation. Denote by $\Psi_x(t, t_0)$ the solution operator for this system with initial point $x \in \mathbb{R}^m$, then

$$D\Phi^{t,t_0}(x) = \Psi_x(t, t_0)\big(D\Phi^{t_0,t_0}(x)\big) = \Psi_x(t, t_0) \cdot \mathbb{1} = \Psi_x(t, t_0). \tag{C.4}$$

This solution operator acts by left-composition on linear maps, so we read off that $\Psi_x(t, t_0) = D\Phi^{t,t_0}(x)$ and find the estimate $\|\Psi_x(t, t_0)\| \leq C_1 \, e^{\rho(t-t_0)}$. Now we turn to the induction step. For $n > 1$, we still have essentially the same solution operator $\Psi_x(t, t_0)$ for the homogeneous part, only now acting by composition on multilinear maps $D^n \Phi^{t,t_0}(x) \in \mathcal{L}^n(\mathbb{R}^m)$: the solution operator is not influenced by considering multilinear maps, as Dv and Ψ act by linear composition from the left, essentially on tangent vectors. Therefore, the same growth estimate for $\Psi_x(t, t_0)$ still holds.

The inhomogeneous terms in (C.3) depend only on the $D^i \Phi^t$, $i < n$ and by the induction hypothesis we can estimate $\|D^i \Phi^t\| \leq C_i \, e^{i \rho t}$. Using variation of constants, the solution can now be written as

$$D^n \Phi^t(x) = \int_{t_0}^{t} \Psi_x(t, \tau) \cdot \sum_{l=2}^{n} D^l v \circ \Phi^\tau \cdot P_{l,n}\big(D^1 \Phi^\tau, \ldots, D^{n-1} \Phi^\tau\big) \, d\tau, \tag{C.5}$$

where the homogeneous part of the solution is zero because $D^n \Phi^{t_0,t_0}(x) = 0$ for $n > 1$. Given that the weighted degree of $P_{l,n}$ is n, we can directly estimate

$$\begin{aligned}\left\|\mathrm{D}^n \Phi^t(x)\right\| &\leq \int_{t_0}^{t} \|\Psi_x(t,\tau)\| \sum_{l=2}^{n} \|\mathrm{D}^l v\| \, \|P_{l,n}\big(\mathrm{D}^1\Phi^\tau, \ldots, \mathrm{D}^{n-1}\Phi^\tau\big)\| \,\mathrm{d}\tau \\ &\leq \int_{t_0}^{t} C_1\, e^{\rho(t-\tau)}\, V\, R\big(\{C_i\}_{i<n}\big)\, e^{n\,\rho(\tau-t_0)}\,\mathrm{d}\tau \\ &= C_1\, V\, R\big(\{C_i\}_{i<n}\big)\, \frac{e^{n\,\rho(t-t_0)} - e^{\rho(t-t_0)}}{(n-1)\,\rho}. \end{aligned} \tag{C.6}$$

The bound R depends on finite sums and products of finite terms, so is finite again. When $\rho > 0$, the denominator is positive and the numerator can be estimated by $e^{n\,\rho(t-t_0)}$; when $\rho < 0$, the numerator can be estimated by $e^{\rho(t-t_0)}$, adding a minus sign to both parts of the fraction. Thus, in both cases (C.1) holds. This completes the induction step. □

Before generalizing this lemma to Riemannian manifolds, we first refine some previous notation. Instead of $\mathbb{R}^m$, we more generally consider linear spaces V, W and spaces $\mathcal{L}^k(V; W)$ of (multi)linear maps for the (higher order) derivatives of maps $f: V \to W$.

Definition C.2 (Homogeneous weighted polynomial) *Let $P_{a,b}(y_1, \ldots, y_n)$ be a polynomial in the variables y_1 to y_n. We call P a homogeneous weighted polynomial of degree (a, b) if it is a homogeneous polynomial of degree a and moreover, each term $y_1^{p_1} \ldots y_n^{p_n}$ has weighted degree*

$$\sum_{i=1}^{n} i \cdot p_i = b. \tag{C.7}$$

As a consequence, such a polynomial cannot have factors y_n for $n > b$ and the factor y_b can only occur as a term on itself when $a = 1$.

This definition can now be used to denote the higher derivatives of a composition of two functions f, g on vector spaces.

Proposition C.3 (Higher order derivatives of compositions of functions) *Let the mapping $x \mapsto f(g(x), x)$ be given with $f: V \times U \to W$ and $g: U \to V$ two sufficiently differentiable functions between vector spaces U, V, W. Then the k-th order derivative of this mapping with respect to x is of the form*

$$\left(\frac{d}{dx}\right)^k f(g(x), x) = \sum_{\substack{l,m\geq 0\\ l+m\leq k\\ (l,m)\neq(0,0)}} \mathrm{D}_1^l \mathrm{D}_2^m f(g(x), x) \cdot P_{l,k-m}\big(\mathrm{D}^1 g(x), \ldots, \mathrm{D}^{k-m} g(x)\big), \tag{C.8}$$

where $P_{l,k-m}$ is a homogeneous weighted polynomial of degree $(l, k-m)$ with l higher order derivatives $\mathrm{D}^i g(x)$ in each term, and weighted degree $k-m$: the total number of derivatives that either produced an additional $\mathrm{D}g(x)$ term or differentiated an existing one.

Remark C.4 We will shorten the notation $P_{l,k}\big(\mathrm{D}^1 g(x), \ldots, \mathrm{D}^k g(x)\big) = P_{l,k}\big(\mathrm{D}^\bullet g(x)\big)$.

Remark C.5 Note that $\mathrm{D}_1^l \mathrm{D}_2^m f(g(x), x)$ is actually an element of the tensor product space $W \otimes \big(V^*\big)^{\otimes l} \otimes \big(U^*\big)^{\otimes m}$ and $P_{l,k-m}$ an element of the $(l, k-m)$-linear maps $V^{\otimes l} \otimes \big(U^*\big)^{\otimes k-m}$, or $(l, k-m)$ tensors, so the composition is indeed a mapping in $W \otimes \big(U^*\big)^{\otimes k} = \mathcal{L}^k(U; W)$, as expected. ◇

Proof This is easily proven by induction. For $k = 1$ we have

$$\frac{\mathrm{d}}{\mathrm{d}x} f(g(x), x) = \mathrm{D}_1 f(g(x), x) \cdot \mathrm{D}g(x) + \mathrm{D}_2 f(g(x), x),$$

which satisfies (C.8). For the induction step we have

$$\begin{aligned}
\left(\frac{\mathrm{d}}{\mathrm{d}x}\right)^{k+1} f(g(x), x) &= \frac{\mathrm{d}}{\mathrm{d}x} \sum_{\substack{l,m\geq 0\\ l+m\leq k\\ (l,m)\neq(0,0)}} \mathrm{D}_1^l \mathrm{D}_2^m f(g(x), x) \cdot P_{l,k-m}\big(\mathrm{D}^\bullet g(x)\big) \\
&= \sum_{\substack{l,m\geq 0\\ l+m\leq k\\ (l,m)\neq(0,0)}} \Big[\quad \mathrm{D}_1^{l+1} \mathrm{D}_2^m f(g(x), x) \cdot \mathrm{D}^1 g(x) \cdot P_{l,k-m}\big(\mathrm{D}^\bullet g(x)\big) \\
&\qquad + \mathrm{D}_1^l \mathrm{D}_2^{m+1} f(g(x), x) \cdot P_{l,k-m}\big(\mathrm{D}^\bullet g(x)\big) \\
&\qquad + \mathrm{D}_1^l \mathrm{D}_2^m f(g(x), x) \cdot \frac{\mathrm{d}}{\mathrm{d}x} P_{l,k-m}\big(\mathrm{D}^\bullet g(x)\big)\Big] \\
&= \sum_{\substack{l,m\geq 0\\ l+m\leq k\\ (l,m)\neq(0,0)}} \Big[\quad \mathrm{D}_1^{l+1} \mathrm{D}_2^m f(g(x), x) \cdot P_{l+1,k+1-m}\big(\mathrm{D}^\bullet g(x)\big) \\
&\qquad + \mathrm{D}_1^l \mathrm{D}_2^{m+1} f(g(x), x) \cdot P_{l,k+m}\big(\mathrm{D}^\bullet g(x)\big) \\
&\qquad + \mathrm{D}_1^l \mathrm{D}_2^m f(g(x), x) \cdot P_{l,k+1-m}\big(\mathrm{D}^\bullet g(x)\big)\Big] \\
&= \sum_{\substack{l,m\geq 0\\ l+m\leq k+1\\ (l,m)\neq(0,0)}} \mathrm{D}_1^l \mathrm{D}_2^m f(g(x), x) \cdot P_{l,k+1-m}\big(\mathrm{D}^\bullet g(x)\big).
\end{aligned}$$

This is again of the form (C.8): $k - m = (k + 1) - (m + 1)$, so all terms can be absorbed in the new sum for $k + 1$. □

Let us make a few remarks on the form of (C.8). The $P_{0,m}$ for $m < k$ are zero, because then we have too few derivatives with respect to x; we have $P_{0,k} = 1$ though. After Definition C.2 it was already noted that in a polynomial of weighted degree k, the factor y_k can only occur as a term on itself, up to a constant factor. More specifically in this case, $\mathrm{D}^k g(x)$ occurs exactly once, in the term

$$D_1 f(g(x), x) \cdot D^k g(x).$$

This can easily be seen by direct calculation or induction. Finally, when the composition mapping is of the form $x \mapsto f(g(x))$, then we only have terms with $m = 0$ and all polynomials in (C.8) have weighted degree k in that case.

The next step is to generalize Lemma C.1 to a Riemannian manifold (M, g). Here we first need to define what we mean by higher derivatives of the flow. The tangent flow $D\Phi^t$ is well-defined as a mapping on TM, but higher derivatives live on higher order tangent bundles $T^k M$. These abstract bundles make doing explicit estimates as in the proof of Lemma C.1 difficult. Instead, we reuse the idea of Definition 2.9 and introduce a different representation of higher derivatives in terms of normal coordinate charts.

Definition C.6 (Higher derivative on Riemannian manifolds) *Let M, N be Riemannian manifolds and $f : M \to N$ a smooth map. With the notation $f_x = \exp_{f(x)}^{-1} \circ f \circ \exp_x$ of f represented in normal coordinate charts, we define for $k \geq 1$ and $x \in M$ the higher order derivative*

$$D^k f(x) = D^k f_x(0) = D^k\big[\exp_{f(x)}^{-1} \circ f \circ \exp_x\big](0) \tag{C.9}$$

as an element of $\mathcal{L}^k(T_x M; T_{f(x)} N)$.

Remark C.7 Definition C.6 can be viewed as creating a more explicit representation of the jet bundle of the trivial fiber bundle $\pi : M \times N \to M$. A map $f : M \to N$ is a section of this trivial bundle and the k-jet of f at a point x is fixed in terms of the derivatives in (C.9) up to order k, in the normal coordinate chart centered at x. We shall see below that this representation is still a (global) bundle, while the explicit choice of normal coordinate charts introduces a convenient norm to measure the jets. ◇

Let us make a few remarks on this choice of representation of higher derivatives. First of all, for $k = 1$ this definition coincides with the ordinary tangent map, as $D\exp_x(0) = \mathbb{1}_{T_x M}$ by the natural identification $T_0(T_x M) \cong T_x M$. Furthermore, this representation of derivatives admits operator norms, and all this behaves nicely under composition of maps by virtue of the property $(f \circ g)_x = f_{g(x)} \circ g_x$ for local coordinate charts:

$$\begin{aligned}
\left\| D^2(f \circ g)(x) \right\| &= \left\| D^2\big[f_{g(x)} \circ g_x\big](0) \right\| \\
&= \left\| D^2 f_{g(x)}(0)\big(Dg_x(0), Dg_x(0)\big) + D f_{g(x)}(0)\big(D^2 g_x(0)\big) \right\| \\
&= \left\| D^2 f(g(x)) \cdot Dg(x)^{\otimes 2} + D f(g(x)) \cdot D^2 g(x) \right\| \\
&\leq \left\| D^2 f(g(x)) \right\| \cdot \|Dg(x)\|^2 + \|D f(g(x))\| \cdot \left\| D^2 g(x) \right\|,
\end{aligned}$$

that is, these operator norms as defined via normal coordinate charts are truly norms and satisfy the usual product rules for compositions of (multi)linear maps.

The operator norms are induced by the norms on the tangent spaces of $\mathrm{T}M$, which in turn are induced by the metric. These norms depend smoothly on the base point, so they glue together to a smooth function $\|\cdot\|\colon \mathrm{T}M \to \mathbb{R}_{\geq 0}$ that we will call a 'bundle norm' on the tangent bundle[4] or sometimes refer to as just a norm on $\mathrm{T}M$. Higher derivatives can be viewed as partial sections of the vector bundle

$$\mathcal{L}^k(\mathrm{T}M; \mathrm{T}N) = \mathrm{T}N \boxtimes (\mathrm{T}M^*)^{\otimes k}. \tag{C.10}$$

That is, we define $\mathcal{L}^k(\mathrm{T}M; \mathrm{T}N)$ as a bundle over $M \times N$ with fiber $\mathcal{L}^k(\mathrm{T}_x M; \mathrm{T}_y N)$ over the point $(x, y) \in M \times N$. This is indicated by the operator $\boxtimes$, which differs from the usual tensor product $\otimes$ in the sense that the new bundle is constructed on the product of the base spaces instead of one common base. Now the k-th order derivative (as in Definition C.6) of a map $f\colon M \to N$ is a section of the bundle (C.10) restricted to the base submanifold $\mathrm{Graph}(f) \subset M \times N$ and the derivative $\mathrm{D}^k f(x)$ is the point in the section over $(x, f(x))$. More generally, we can define vector bundles of (l, k)-linear maps

$$\mathcal{L}^{l,k}(\mathrm{T}M; \mathrm{T}N) = (\mathrm{T}N)^{\otimes l} \boxtimes (\mathrm{T}M^*)^{\otimes k} \tag{C.11}$$

and the disjoint union of all these bundles. The bundle norms on $\mathrm{T}M$ and $\mathrm{T}N$ together naturally induce bundle operator norms on these. From here on, we set $M = N$ and assume that $f = \Phi^t$ is a flow.

To finally generalize Lemma C.1 to Riemannian manifolds, there is still one issue to tackle. When taking the time-derivative as in (C.3), the target base point $\Phi^t(x)$ changes. This suggests that a covariant derivative is required. The $\mathcal{L}^{l,k}(\mathrm{T}M; \mathrm{T}M)$ are smooth manifolds in a natural way, however, so both the tangent vector $\frac{\mathrm{d}}{\mathrm{d}t}\mathrm{D}^k\Phi^t(x)$ and the differential of $\|\cdot\|$ are well defined in this interpretation and independent of a connection, and certainly their product $\frac{\mathrm{d}}{\mathrm{d}t}\left\|\mathrm{D}^k\Phi^t(x)\right\|$ is. The tangent exponential maps $\mathrm{D}\exp$ at x and $\Phi^t(x)$ together induce a local coordinate chart on $\mathcal{L}^{l,k}(\mathrm{T}M; \mathrm{T}M)$ in a neighborhood of $\mathcal{L}^{l,k}(\mathrm{T}_x M; \mathrm{T}_{\Phi^t(x)} M)$. We will use these local coordinates for explicit calculations.

The dependence on the base point of the norms and normal coordinate charts in (C.9) introduces additional terms when formulating equations (C.3) and (C.5) on a Riemannian manifold. Under the assumption that (M, g) is of bounded geometry, however, all these additional terms will be globally bounded. Hence, these will only contribute to the overall constants C_n in Lemma C.1, but not influence the basic result.

Lemma C.8 (Exponential growth estimates on a Riemannian manifold) *Let $\Phi^{t,t_0} \in C^{k\geq 1}$ be the flow of a time-dependent vector field v on a Riemannian manifold*

[4] Note that this is stronger than a Finsler manifold as the Finsler structure $F\colon \mathrm{T}M \to \mathbb{R}_{\geq 0}$ is allowed to be asymmetric, that is, on each tangent space, F need only scale linearly for positive scalars. I did not investigate whether it is possible to generalize this theory to Finsler manifolds.

(M, g) of $(k+3)$-bounded geometry. Let $v(t, \cdot) \in \mathfrak{X}_b^k(M)$ with all derivatives jointly continuous in $(t, x) \in \mathbb{R} \times M$ and uniformly bounded by $V < \infty$ with respect to Definition C.6. Suppose that $\|D\Phi^{t,t_0}(x)\| \leq C_1\, e^{\rho(t-t_0)}$ for all $x \in M$, $t \geq t_0$ and fixed $C_1 > 0$, $\rho \neq 0$. Then for each n, $1 \leq n \leq k$ there exists a bound $C_n > 0$ such that

$$\forall x \in M,\ t \geq t_0\colon \left\|D^n \Phi^{t,t_0}(x)\right\| \leq \begin{cases} C_n\, e^{n\,\rho(t-t_0)} & \text{if } \rho > 0, \\ C_n\, e^{\rho(t-t_0)} & \text{if } \rho < 0. \end{cases} \tag{C.12}$$

Proof The proof is basically the same as the proof of Lemma C.1, with additional technicalities due to M being a manifold. We will focus on these.

Equation (C.3) can be formulated in terms of the tangent normal coordinate chart

$$\mathrm{D}\exp_y^{-1}\colon \mathrm{T}B(y;\delta) \subset \mathrm{T}M \to \mathrm{T}(\mathrm{T}_yM) \cong (\mathrm{T}_yM)^2$$

with $y = \Phi^t(x)$ fixed. Note that we are finally interested in the growth behavior of $t \mapsto \|D^n\Phi^t(x)\|$; this is defined in a coordinate-free way, so it is not influenced by our choice of intermediate coordinates. In these normal coordinates, both the metric and its derivatives are bounded due to Theorem 2.4, and the vector field is C^k bounded by assumption. We have

$$\begin{aligned}
&\frac{\mathrm{d}}{\mathrm{d}t}\mathrm{D}\exp_y^{-1} \circ \mathrm{D}^n\Phi^t(x) \\
&\quad= \frac{\mathrm{d}}{\mathrm{d}t}\mathrm{D}\exp_y^{-1} \circ \mathrm{D}^n\big[\exp_{\Phi^t(x)}^{-1} \circ \Phi^t \circ \exp_x\big](0) \\
&\quad= \frac{\mathrm{d}}{\mathrm{d}t}\mathrm{D}\exp_y^{-1} \circ \sum_{l=1}^{n} \mathrm{D}^l\big[\exp_{\Phi^t(x)}^{-1} \circ \exp_y\big](0) \cdot P_{l,n}\big(\mathrm{D}^\bullet\big[\exp_y^{-1} \circ \Phi^t \circ \exp_x\big](0)\big).
\end{aligned}$$

This splits the dependence on t in the target base point $\Phi^t(x)$ from that in the derivatives $\mathrm{D}^n\Phi^t$ itself. Note that the sum must be interpreted as a sum of terms in the single fiber $\mathcal{L}^n(\mathrm{T}_xM; \mathrm{T}_yM)$ over the base point (x, y). By using the coordinate map $\mathrm{D}\exp_y^{-1}$, we transferred the problem to fixed linear spaces, which allows us to make sense of the differentiation with respect to t. In other words, $\mathrm{D}\exp_y^{-1}$ induces locally trivializing coordinates for $\mathcal{L}^n(\mathrm{T}_xM; \mathrm{T}M)$ in a neighborhood of y with x fixed. As $\mathrm{D}\exp_y^{-1}$ is linear on the fibers, we can distribute it over the sum to further obtain

$$\begin{aligned}
&= \sum_{l=1}^{n} \frac{\mathrm{d}}{\mathrm{d}t}\Big[\mathrm{D}\exp_y^{-1} \circ \mathrm{D}^l\big[\exp_{\Phi^t(x)}^{-1} \circ \exp_y\big](0) \cdot P_{l,n}\big(\mathrm{D}^\bullet\big[\exp_y^{-1} \circ \Phi^t \circ \exp_x\big](0)\big)\Big] \\
&= \frac{\mathrm{d}}{\mathrm{d}t}\Big[\mathrm{D}\exp_y^{-1} \circ \mathrm{D}\big[\exp_{\Phi^t(x)}^{-1} \circ \exp_y\big](0) \cdot \mathrm{D}^n\big[\exp_y^{-1} \circ \Phi^t \circ \exp_x\big](0)\Big] + \sum_{l=2}^{n} \frac{\mathrm{d}}{\mathrm{d}t}[\ldots].
\end{aligned} \tag{C.13}$$

In the last line, the homogeneous part is separated from the non-homogeneous terms as in (C.3).

Working out the details of the homogeneous part, we obtain[5]

$$
\begin{aligned}
&\frac{\mathrm{d}}{\mathrm{d}t}\Big[\mathrm{D}\exp_y^{-1}\circ\mathrm{D}\big[\exp_{\Phi^t(x)}^{-1}\circ\exp_y\big](0)\cdot\mathrm{D}^n\big[\exp_y^{-1}\circ\Phi^t\circ\exp_x\big](0)\Big]\\
&=\frac{\mathrm{d}}{\mathrm{d}t}\Big[\mathrm{D}\exp_y^{-1}\circ\mathrm{D}\big[\exp_{\Phi^t(x)}^{-1}\circ\exp_y\big](0)\Big]\cdot\mathrm{D}^n\Phi^t(x)\\
&\quad+\mathrm{D}\big[\mathrm{D}\exp_y^{-1}\big]\cdot\mathrm{D}^n\frac{\mathrm{d}}{\mathrm{d}t}\big[\exp_y^{-1}\circ\Phi^t\circ\exp_x\big](0)\\
&=\frac{\mathrm{d}}{\mathrm{d}t}\Big[\mathrm{D}\exp_y^{-1}\circ\mathrm{D}\big[\exp_{\Phi^t(x)}^{-1}\circ\exp_y\big](0)\Big]\cdot\mathrm{D}^n\Phi^t(x)\\
&\quad+\mathrm{D}\big[\mathrm{D}\exp_y^{-1}\big]\cdot\sum_{l=1}^{n}\mathrm{D}^l\big[\mathrm{D}\exp_y^{-1}\circ v\circ\exp_y\big](0)\cdot P_{l,n}(\mathrm{D}^\bullet\Phi^t(x)).
\end{aligned}
$$

Note that again all terms $l \geq 2$ in the sum are inhomogeneous terms that we will add to those already present in (C.13). The homogeneous term is some linear vector field acting (from the left) on $\mathrm{D}^n\Phi^t(x)$ and it is precisely the vector field generating $\mathrm{D}\Phi^t(x)$, which is the original case $n = 1$. Hence, we can again define the operator Ψ_x^{t,t_0} as post-composition with $\mathrm{D}\Phi^t(x)$ and write the flow of $\mathrm{D}^n\Phi^t(x)$ using a variation of constants integral with all the non-homogeneous terms. These terms again contain only lower order derivative flows $\mathrm{D}^l\Phi^t(x),\ l < n$.

We can now take the operator norm of this expression. In principle we should be careful that this bundle norm depends on the changing target point $\Phi^t(x)$. The normal coordinates were chosen around $y = \Phi^t(x)$, however, and in these coordinates the derivative of the metric at the origin (corresponding to y) is zero, hence the norm has zero derivative. We can thus simply apply the operator norm to the variation of constants integral and obtain estimates as in (C.6). The additional factors introduced by differentiation of normal coordinate transition maps are bounded by Lemma 2.6 under the assumption that (M, g) is of (k+3)-bounded geometry. The inhomogeneous terms still contain at least one factor $\mathrm{D}^l\Phi^t(x)$, so the result in case $\rho < 0$ holds as well. □

These exponential growth results can be extended further to uniform and Hölder continuity in the highest derivatives. The Hölder continuity then is with respect to the growth rate $(k + \alpha)\rho$, where k is the order of the derivative and $0 < \alpha \leq 1$ the Hölder constant. Thus, α-Hölder continuity can be viewed as a fractional derivative; Lipschitz continuity (when $\alpha = 1$) can indeed be viewed as almost differentiability to one higher order. The case $\alpha = 0$ we shall identify with uniform continuity. Here we

---

[5] The time derivative of $\mathrm{D}\exp_y^{-1}\circ\mathrm{D}\big[\exp_{\Phi^t(x)}^{-1}\circ\exp_y\big](0)$ actually turns out to be zero in local coordinates. This follows from an analysis of the exponential map as the time-one geodesic flow in normal coordinates around y. This result is not relevant for us, so we leave out this tedious calculation.

have no explicit modulus of continuity, which requires an arbitrarily small additional $\mu > 0$ in the exponent $k\rho + \mu$ to compensate.

Remark C.9 (*On using a global continuity modulus*) In the next lemma, as well as in Corollary C.12 below, we shall make abuse of notation in writing expressions such as $\|s(x_2) - s(x_1)\|$, where s is a section of a vector bundle, cf. (C.15), that is, we compare objects that live in different fibers of a vector bundle.[6] This notation should be interpreted according to Remark 2.12. That is, if x_1, x_2 are M-close in the spirit of Definition 2.8, then this is well-defined in terms of local charts, and for continuity estimates this is equivalent to an estimate by identification of the vector bundle over x_1, x_2 via parallel transport, cf. Proposition 2.13. If x_1 and x_2 are not close, then we can use any choice of isometric identification of the vector bundle over these points, such as the construction of parallel transport along solutions curves in Sect. 3.7.4. In this case the notation can effectively be interpreted as an estimation by the sum of the norms of the separate terms with the triangle inequality. When applying this lemma, we shall always have such an isometric identification at hand, hence these arguments can be made rigorous, and the notation provides a sensible heuristic then. ◇

Lemma C.10 (Exponential growth estimates with Hölder continuity) *Let $\Phi^{t,t_0} \in C^{k\geq 1}$ be the flow of a time-dependent vector field v on a Riemannian manifold (M, g) of $(k+3)$-bounded geometry. Let D denote the partial derivative with respect to the spatial variable $x \in M$ as in Definition C.6 and let $v(t, \cdot) \in \mathfrak{X}^{k,\alpha}_{b,u}(M)$, $0 < \alpha \leq 1$ with all derivatives jointly continuous in $(t, x) \in \mathbb{R} \times M$. Suppose that $\|\mathrm{D}\Phi^{t,t_0}(x)\| \leq C_1\, e^{\rho(t-t_0)}$ for all $x \in M$, $t \geq t_0$ and fixed $C_1 > 0$, $\rho > 0$.*

Then in addition to the results of Lemma C.8, there exists a bound $C_{k,\alpha} > 0$ such that

$$\forall\, t \geq t_0 \colon \left\|\mathrm{D}^k \Phi^{t,t_0}\right\|_\alpha \leq C_{k,\alpha}\, e^{(k+\alpha)\,\rho(t-t_0)}. \tag{C.14}$$

If instead $v(t, \cdot) \in \mathfrak{X}^k_{b,u}(M)$, i.e. the special case $\alpha = 0$, then for each $\mu > 0$ there exists a continuity modulus $\varepsilon_{k,\mu}$ such that

$$\forall\, t \geq t_0 \colon \left\|\mathrm{D}^k \Phi^{t,t_0}(x_2) - \mathrm{D}^k \Phi^{t,t_0}(x_1)\right\| \leq \varepsilon_{k,\mu}(d(x_1, x_2))\, e^{(k\,\rho+\mu)(t-t_0)}, \tag{C.15}$$

that is, $x \mapsto \big(t \mapsto \mathrm{D}^k \Phi^{t,t_0}(x)\big)$ is uniformly continuous in x, in $\|\cdot\|_{k\,\rho+\mu}$-norm.

Remark C.11 We restricted this lemma to the case $\rho > 0$ only. A result similar to that in C.8 for $\rho < 0$ could be obtained for completeness sake, but it clutters the already detailed proof, while we do not need the result. ◇

---

[6] Note that the higher derivatives $\mathrm{D}^k \Phi^t(x)$ of a flow are actually interpreted as elements of a bundle of type (C.10). These bundles are still naturally induced by the tangent bundles of underlying manifolds, so all bounded geometry techniques, such as uniformity of normal coordinate charts, unique local trivializations by parallel transport, are induced on these bundles as well.

Proof The idea of the proof is essentially the same as that of Lemma C.8. The additional difficulty is that (Hölder) continuity requires finite, non-differential estimates when comparing any two flows starting from different initial points $x_1,\ x_2 \in M$.

Let $d(x_1, x_2) < \delta_M$ where δ_M is M-small as in Definition 2.8. We drop t_0 from the notation and define $\xi_i(t) = \Phi^t(x_i),\ i = 1, 2$ as the solution curves with x_i as initial conditions. We want to study the growth behavior of

$$t \mapsto \mathrm{D}^k \Phi^t(x_2) - \mathrm{D}^k \Phi^t(x_1). \tag{C.16}$$

Note that this difference is defined with respect to coordinate charts at source and target that contain $x_1,\ x_2$ and $\xi_1(t),\ \xi_2(t)$, respectively, but not in general.

We denote by γ_t the unique shortest geodesic that connects $\xi_1(t)$ to $\xi_2(t)$ when $d(\xi_1(t), \xi_2(t)) < \delta_M$. Next, we set

$$\Upsilon^t = \mathrm{D}^k \Phi^t(x_2) \cdot \Pi(\gamma_0)^{\otimes k} - \Pi(\gamma_t) \cdot \mathrm{D}^k \Phi^t(x_1) \in \mathcal{L}^k\big(\mathrm{T}_{x_1} M; \mathrm{T}_{\xi_2(t)} M\big) \tag{C.17}$$

to be the difference of the respective k-th order derivative flows, parallel transported to matching spaces at their source and target. It is easily verified that Υ^t satisfies initial conditions $\Upsilon^{t_0, t_0} = 0$ for any $k \geq 1$.

Due to Proposition 2.13, the formulation in (C.17) with parallel transport to measure variation of the flows is equivalent to measuring (C.16) in normal coordinate charts. Hence, if we study $\|\Upsilon^t\|$ in charts, we may drop[7] the parallel transport terms at the cost of an (unimportant) global factor in the estimates. We assume that $d(\xi_1(t), \xi_2(t)) < \delta_M$ and study Υ^t in a normal coordinate chart covering both points. Taking the difference of (C.3) with $x_1,\ x_2$ inserted, we see that Υ^t satisfies the differential equation

$$\begin{aligned}\frac{\mathrm{d}}{\mathrm{d}t}\Upsilon^t &= \mathrm{D}v \circ \Phi^t(x_2) \cdot \Upsilon^t + \big[\mathrm{D}v \circ \Phi^t(x_2) - \mathrm{D}v \circ \Phi^t(x_1)\big] \cdot \mathrm{D}^k \Phi^t(x_1) \\ &\quad + \sum_{l=2}^{k} \mathrm{D}^l v \circ \Phi^t(x_2) \cdot P_{l,k}\big(\mathrm{D}^\bullet \Phi^t(x_2)\big) - (x_2 \rightsquigarrow x_1).\end{aligned}$$

This equation provides a variation of constants integral for Υ^t based on the flow $\Psi_{x_2}(t, t_0)$:

---

7 We could include the parallel transport terms, repeat similar arguments as in the proof of Lemma C.8 and express everything in (induced) normal coordinate charts, but this would clutter the proof here even more. These terms would all be bounded and Lipschitz continuous by bounded geometry, hence not essentially alter the result.

$$
\Upsilon^t = \int_{t_0}^{t} \Psi_{x_2}(t,\tau)\cdot\big[\mathrm{D}v\circ\Phi^\tau(x_2) - \mathrm{D}v\circ\Phi^\tau(x_1)\big]\cdot \mathrm{D}^k\Phi^\tau(x_1)
$$

$$
+ \Psi_{x_2}(t,\tau)\cdot\left[\sum_{l=2}^{k} \mathrm{D}^l v\circ\Phi^\tau(x_2)\cdot P_{l,k}\big(\mathrm{D}^\bullet\Phi^\tau(x_2)\big) - (x_2 \rightsquigarrow x_1)\right] \mathrm{d}\tau. \tag{C.18}
$$

We proceed by induction over k. For $k = 1$ we only have the first term of the integrand. Using that $\mathrm{D}v$ is uniformly α-Hölder, we have

$$
\begin{aligned}
\|\Upsilon^t\| &\le \int_{t_0}^{t} \|\Psi_{x_2}(t,\tau)\| \, \|\mathrm{D}v\circ\Phi^\tau(x_2) - \mathrm{D}v\circ\Phi^\tau(x_1)\| \, \|\mathrm{D}\Phi^\tau(x_1)\| \,\mathrm{d}\tau \\
&\le \int_{t_0}^{t} C_1\, e^{\rho(t-\tau)}\, \|\mathrm{D}v\|_\alpha \, \|\Phi^\tau(x_2) - \Phi^\tau(x_1)\|^\alpha \, C_1\, e^{\rho(\tau-t_0)}\,\mathrm{d}\tau \\
&\le C_1^2\, \|\mathrm{D}v\|_\alpha\, e^{\rho(t-\tau)} \int_{t_0}^{t} \big(C_1\, e^{\rho(\tau-t_0)}\, \|x_2 - x_1\|\big)^\alpha \,\mathrm{d}\tau \\
&\le C_1^{2+\alpha}\, \|\mathrm{D}v\|_\alpha\, e^{\rho(t-\tau)}\, \|x_2 - x_1\|^\alpha\, \frac{e^{\alpha\,\rho(t-t_0)}}{\alpha\,\rho}.
\end{aligned}
$$

Next, in the induction step for $k > 1$, we get the additional terms from (C.18) in the integrand. These are (up to constants) a product of the flow Ψ, $\mathrm{D}^l v\circ\Phi^\tau(x)$ and $\mathrm{D}^i\Phi(x)$'s with weighted degree k. The terms $\mathrm{D}^l v\circ\Phi^\tau(x)$ are uniformly α-Hölder continuous in x analogous to the case $k = 1$ above. Each of the $\mathrm{D}^i\Phi(x)$'s satisfies the Hölder estimate of this lemma by the induction hypothesis and the growth estimates of Lemma C.1. Hence, for each term in the integrand, we obtain Hölder continuity with respect to x with growth behavior at most $e^{\rho(t-\tau)}\, e^{(k+\alpha)\rho(\tau-t_0)}$. Integration then yields the stated result.

Finally, the uniformly continuous case is an extension along the same lines as Corollary B.3. The map $x \mapsto \mathrm{D}^l v\circ\Phi^\tau(x)$ is uniformly continuous when measured in $\|\cdot\|_\mu$-norm and by induction the alternative result (C.15) follows. □

Finally, we extend the Hölder continuous growth estimates to a parameter dependent version. This is formulated to exactly fit the context of derivatives of T_X with respect to $y \in B^\rho_\eta(I;Y)$, such as in (3.43b). Note that Remark C.9 applies again.

Corollary C.12 (Exponential growth with Hölder continuity and a parameter) *Assume the setting of Lemma C.10. Let the vector field v furthermore depend on a third variable $y \in Y$ such that $\mathrm{D}^l_x v(t,x,\cdot) \in C^\alpha_{b,u}$ for all $0 \le l \le k$, uniformly in t, x and that all original bounds are uniform in y as well. Let $\eta \in B^\rho(\mathbb{R};Y)$ denote a curve in Y and Φ^{t,t_0}_η the flow of $v(t,\cdot,\eta(t))$. Assume that $\eta \mapsto \big(t \mapsto \Phi^{t,t_0}_\eta(x)\big)$ is uniformly Lipschitz with respect to the distance function d_ρ on curves $C(\mathbb{R}_{\ge t_0};X)$.*

Then the map $\eta \mapsto \mathrm{D}^k\Phi_\eta$ is Hölder continuous in the sense that there exists a bound $C_{k,\alpha,Y} > 0$ such that

$$
\forall\, t \ge t_0,\ x \in X:\ \left\|\eta \mapsto \mathrm{D}^k\Phi^{t,t_0}_\eta(x)\right\|_\alpha \le C_{Y,\alpha}\, e^{(k+\alpha)\,\rho(t-t_0)}. \tag{C.19}
$$

In case of uniform continuity (i.e. $\alpha = 0$), then for each $\mu > 0$ there exists a continuity modulus $\varepsilon_{k,\mu,Y}$ such that

$$\forall\, t \geq t_0,\ x \in X : \left\| \mathrm{D}^k \Phi_{\eta_2}^{t,t_0}(x) - \mathrm{D}^k \Phi_{\eta_1}^{t,t_0}(x) \right\| \leq \varepsilon_{k,\mu,Y}(d(\eta_1, \eta_2))\, e^{(k\,\rho+\mu)(t-t_0)}. \tag{C.20}$$

In both cases we interpret the continuity moduli as globally defined using Remark 2.12.

Proof The proof closely follows that of Lemma C.10; let us indicate the differences.

We define the variation

$$\Upsilon^t = \mathrm{D}^k \Phi_{\eta_2}^t(x) \cdot \Pi(\gamma_0)^{\otimes k} - \Pi(\gamma_t) \cdot \mathrm{D}^k \Phi_{\eta_1}^t(x) \tag{C.21}$$

and study it by a variation of constants integral in local charts, similar to (C.18). In this case we obtain

$$\Upsilon^t = \int_{t_0}^{t} \Psi_{\eta_2}(t,\tau) \cdot \Bigg(\Big[\mathrm{D}v\big(\tau, \Phi_{\eta_2}^\tau(x), \eta_2(\tau)\big) - \mathrm{D}v\big(\tau, \Phi_{\eta_1}^\tau(x), \eta_1(\tau)\big) \Big] \cdot \mathrm{D}^k \Phi_{\eta_2}^\tau(x) \\ + \Big[\sum_{l=2}^{k} \mathrm{D}^l v\big(\tau, \Phi_{\eta_2}^\tau(x), \eta_2(\tau)\big) \cdot P_{l,k}\big(\mathrm{D}^\bullet \Phi_{\eta_2}^\tau(x)\big) - (2 \rightsquigarrow 1) \Big] \Bigg)\, \mathrm{d}\tau. \tag{C.22}$$

As in Lemma C10, the factors $\Psi_{\eta_2}(t,\tau)$ and $\mathrm{D}^l \Phi_{\eta_2}^\tau(x)$ satisfy appropriate exponential growth conditions. By induction over $l < k$ the maps $\eta \mapsto \mathrm{D}^l \Phi_\eta^t(x)$ are α-Hölder continuous, while all $\mathrm{D}^l v$ are uniformly α-Hölder in x, y, and $\eta \mapsto \mathrm{D}^l \Phi_\eta^t(x)$ is uniformly Lipschitz by assumption, so $\eta \mapsto \mathrm{D}^l v\big(t, \Phi_\eta^t(x), \eta(t)\big)$ is also α-Hölder when measured in $\|\cdot\|_{\alpha\,\rho}$-norm (or in $\|\cdot\|_\mu$-norm in case of uniform continuity, see Appendix B).

In each term of the integrand, we can estimate the variation with respect to η as a sum of the variations with respect to each factor (a product rule). The factor that is being varied adds $e^{\alpha\,\rho(\tau-t_0)}$ (or $e^{\mu(\tau-t_0)}$ in case $\alpha = 0$) to the overall growth estimate. The proof is completed by inserting all these estimates into (C.22) and again using the fact that we have a finite number of globally bounded terms. □

Appendix D
The Fiber Contraction Theorem

In this appendix, we give a proof of the fiber contraction theorem. This result is originally due to Hirsch and Pugh [HP70]; the proof presented here is taken from Vanderbauwhede [Van89, p. 105]. The fiber contraction theorem is a convenient general tool to obtain convergence of functions in C^k-norm when a direct contraction in C^k-norm is not available. Instead, one inductively constructs contractions for the k-thderivative with all lower order derivatives assumed fixed. If this contraction depends continuously on the lower order derivatives, then the fiber contraction theorem can be applied to conclude that the sequence of the function together with its derivatives converges to a fixed point. With the additional theorem on the differentiability of limit functions, it can then be concluded that the sequence converges in C^k-norm.

Theorem D.1 (Fiber contraction theorem) *Let X be a topological space, (Y, d) a complete metric space and let $F : X \times Y \to X \times Y$ be a fiber mapping, that is, $F(x, y) = \big(F_1(x), F_2(x, y)\big)$, with the following properties:*

1. *F_1 has a unique, globally attracting fixed point $x^\star \in X$, that is,*
$$\forall x \in X\colon \lim_{n\to\infty} F_1^n(x) = x^\star;$$
2. *there is a neighborhood $U \subset X$ of $x^\star$, such that $F_2 : U \times Y \to Y$ is a uniform contraction on Y with contraction factor $q < 1$; let $y^\star \in Y$ denote the unique fixed point of $F_2(x^\star, \cdot) : Y \to Y$, as given by the Banach fixed point theorem;*
3. *the mapping $F_2(\cdot, y^\star)\colon X \to Y$ is continuous.*

If only properties (1) and (2) are assumed, then $(x^\star, y^\star)$ is the unique fixed point of F. If moreover (3) holds, then this fixed point is globally attractive.

Proof The point $(x^\star, y^\star)$ is clearly the unique fixed point of F, where property (1) implies uniqueness of $x^\star$ as fixed point of F_1 and (2) uniqueness of $y^\star$ under $F_2(x^\star, \cdot)$.

The point $x^\star$ is by assumption attractive under F_1, thus for the final conclusion of global attractivity, it remains to show that $y \to y^\star$ under F.

J. Eldering, *Normally Hyperbolic Invariant Manifolds*, Atlantis Series in Dynamical Systems 2, DOI: 10.2991/978-94-6239-003-4,

Let $(x, y) \in X \times Y$ be arbitrary and consider the sequence $(x_n, y_n) = F^n(x, y)$ for $n \geq 0$. Since $x_n \to x^* \in U$, there exists an $N \in \mathbb{N}$ such that $x_n \in U$ for all $n \geq N$. By shifting the sequence (x_n, y_n), we can assume without loss of generality that $x_n \in U$ for all $n \geq 0$ and use property (2) to estimate

$$\begin{aligned} d(y_{n+1}, y^\star) &= d(F_2(x_n, y_n), F_2(x^\star, y^\star)) \\ &\leq d(F_2(x_n, y_n), F_2(x_n, y^\star)) + d(F_2(x_n, y^\star), F_2(x^\star, y^\star)) \\ &\leq q\, d(y_n, y^\star) + \alpha_n. \end{aligned} \tag{D.1}$$

On the other hand, $\alpha_n = d(F_2(x_n, y^\star), F_2(x^\star, y^\star)) \to 0$ as $n \to \infty$ from properties (1) and (3). Let $\bar{\alpha}_k = \sup_{n \geq k} \alpha_n$, then we also have $\bar{\alpha}_k \to 0$.

For each $k \in \mathbb{N}$, let $\delta_{k,k} = d(y_k, y^\star)$ and recursively define $\delta_{n+1,k} = q\, \delta_{n,k} + \bar{\alpha}_k$. From (D.1) we see that $d(y_n, y^\star) \leq \delta_{n,k}$ when $n \geq k$. Now the map $f : \delta \mapsto q\,\delta + \bar{\alpha}$ is a contraction for any $\bar{\alpha} \in \mathbb{R}$, so it has a unique, attractive fixed point $\delta^\star(\bar{\alpha})$ and solving the equation $f(\delta^\star) = \delta^\star$ yields

$$\delta^\star = \frac{\bar{\alpha}}{1-q}.$$

Let $\varepsilon > 0$ be given and choose k large enough that $\bar{\alpha}_k < \frac{1}{2}(1-q)\varepsilon$. As $\lim_{n\to\infty} \delta_{n,k} = \delta_k^\star$ we see that there exists some N such that

$$\forall\, n \geq N: \quad \delta_{n,k} < 2\delta_k^\star = \frac{2\,\bar{\alpha}_k}{1-q} < \varepsilon.$$

From this we conclude that $d(y_n, y^\star) < \varepsilon$ for all $n \geq N$. □

The following theorem is quite standard. We shall extend it to smooth manifolds and higher derivatives, though.

Theorem D.2 (Differentiability of limit functions) *Let Y be a Banach space and let $C^k(\mathbb{R}^n; Y)$ denote the space of C^k functions $\mathbb{R}^n \to Y$ equipped with the weak Whitney topology. Let $\{f_n\}_{n\geq 0}$ be a sequence in $C^1(\mathbb{R}^n; Y)$ that converges to $f \in C^0(\mathbb{R}^n; Y)$ with respect to the C^0 topology, and assume that there is a function $g \in C^0(\mathbb{R}^n; \mathcal{L}(\mathbb{R}^n; Y))$ such that $\mathrm{D} f_n \to g$.*

Then $\mathrm{D} f = g$, or in other words, $f_n \to f$ in $C^1(\mathbb{R}^n; Y)$ with respect to the weak Whitney topology.

Proof By the fundamental theorem of calculus we have

$$f_n(x + t\,h) = f_n(x) + \int_0^t \frac{\mathrm{d}}{\mathrm{d}\tau} f_n(x + \tau\,h)\, \mathrm{d}\tau = f_n(x) + \int_0^t \mathrm{D} f_n(x + \tau\,h) \cdot h\, \mathrm{d}\tau.$$

Uniform convergence of $\mathrm{D} f_n \to g$ on the compact set $\{x + \tau h | \tau \in [0, t]\}$ allows us to take the limit $n \to \infty$ inside the integral to obtain

$$f(x + t\,h) = f(x) + \int_0^t g(x + \tau\,h) \cdot h \,\mathrm{d}\tau,$$

and by differentiation with respect to t we conclude that $g(x) \cdot h$ is the directional derivative of f at x along h.

Note that $g(x)\colon \mathbb{R}^n \to Y$ is a bounded linear operator by assumption, so let us verify that it is the total derivative, $\mathrm{D}f(x) = g(x)$, that is,

$$\lim_{h \to 0} \frac{\|f(x+h) - f(x) - g(x) \cdot h\|}{\|h\|} = 0.$$

Using the mean value theorem, we have

$$\|f(x+h) - f(x) - g(x) \cdot h\| \leq \sup_{\xi \in [0,1]} \|g(x + \xi\,h) - g(x)\| \|h\|$$

and g is continuous, so indeed differentiability holds and $\mathrm{D}f(x) = g(x)$. □

Remark D.3 The statement that f is differentiable at x is local, so this result immediately translates to maps $C^1(X; Y)$ with X a smooth manifold by considering a local coordinate chart around $x \in X$.

This theorem could probably be generalized even further such that X, Y are allowed to be Banach manifolds. The fact that g is continuous linear by assumption mitigates possible convergence problems when having to consider infinitely many independent partial derivatives. We should be careful though, since the weak Whitney (or compact-open) topology is not clearly defined anymore when X is infinite-dimensional. ◇

Corollary D.4 *Assume the setting of Theorem D.2. Let $\{f_n\}_{n \geq 0}$ be a sequence in $C^{k \geq 2}(\mathbb{R}^n; Y)$ that converges to f in $C^{k-1}(\mathbb{R}^n; Y)$ and let $\mathrm{D}^k f_n \to g$ converge in $C^0(\mathbb{R}^n; \mathcal{L}^k(\mathbb{R}^n; Y))$. Then $f_n \to f$ converges in $C^k(\mathbb{R}^n; Y)$.*

This is a trivial extension of Theorem D.2 when using the natural identification $\mathcal{L}(\mathbb{R}^n; \mathcal{L}^{k-1}(\mathbb{R}^n; Y)) \cong \mathcal{L}^k(\mathbb{R}^n; Y)$.

Appendix E
Nonlinear Variation of Flows

In this appendix we collect two results on variation of nonlinear flows. The first is a generalization of Lagrange's variation of constants formula and the second is an application of it to calculate the derivative of a flow with respect to parameters. Both results are formulated for fully nonlinear flows.

The classical variation of constants integral due to Lagrange is well known. Although Lagrange applied this method to the nonlinear problem of orbital mechanics, a less known result of Alekseev [Ale61] (see also [LL69, p. 78]) generalizes the variation of constants integral to the full nonlinear case.

Theorem E.1 (Nonlinear variation of constants) *Let X be a smooth manifold and let $\Phi^{t,t_0}(x)$ be the flow generated by the time-dependent vector field $v(t,x)$, locally Lipschitz in x. Let $r(t,x)$ be an arbitrary (not necessarily small) perturbation, locally Lipschitz in x as well. Then $\Phi_r^{t,t_0}(x)$ is the flow generated by $v+r$ if and only if it satisfies the nonlinear variation of constants formula*

$$\Phi_r^{t,t_0}(x) = \Phi^{t,t_0}(x) + \int_{t_0}^{t} \mathrm{D}\Phi(t,\tau,\Phi_r^{\tau,t_0}(x))\, r(\tau,\Phi_r^{\tau,t_0}(x))\, d\tau. \tag{E.1}$$

Proof Using uniqueness of solutions, it is sufficient to show that (E.1) satisfies the differential equation and initial conditions $\Phi_r^{t,t}(x) = x$. The latter follows automatically from $\Phi^{t,t}(x) = x$. For the first part, we differentiate

$$\begin{aligned}
\frac{\mathrm{d}}{\mathrm{d}\tau}\Big[\Phi^{t,\tau} \circ \Phi_r^{\tau,t_0}(x)\Big] &= \frac{\partial}{\partial\tau}\, \Phi^{t,\tau}(y)\Big|_{y=\Phi_r^{\tau,t_0}(x)} + \mathrm{D}\Phi^{t,\tau}(\Phi_r^{\tau,t_0}(x)) \cdot \frac{\mathrm{d}}{\mathrm{d}\tau}\Phi_r^{\tau,t_0}(x) \\
&= -\mathrm{D}\Phi^{t,\tau}(\Phi_r^{\tau,t_0}(x)) \cdot v(\tau,\Phi_r^{\tau,t_0}(x)) \\
&\quad + \mathrm{D}\Phi^{t,\tau}(\Phi_r^{\tau,t_0}(x)) \cdot (v+r)(\tau,\Phi_r^{\tau,t_0}(x)) \\
&= \mathrm{D}\Phi^{t,\tau}(\Phi_r^{\tau,t_0}(x)) \cdot r(\tau,\Phi_r^{\tau,t_0}(x)).
\end{aligned}$$

This expression yields (E.1) when integrated from t_0 to t. □

J. Eldering, *Normally Hyperbolic Invariant Manifolds*, Atlantis Series in Dynamical Systems 2, DOI: 10.2991/978-94-6239-003-4,

Notice that (E.1) looks ill-defined on a manifold, but should be read as integration from the point x along the vector field defined by the integrand, which is indeed, for each $\tau \in [t_0, t]$, exactly defined to be the tangent vector to the curve $\tau \mapsto \Phi^{t,\tau} \circ \Phi_r^{\tau,t_0}(x)$, making the equation self-consistent. If (X, g) is a Riemannian manifold, then this formula yields the distance estimate

$$d\big(\Phi_r^{t,t_0}(x), \Phi^{t,t_0}(x)\big) \leq \int_{t_0}^{t} \big\| \mathrm{D}\Phi(t, \tau, \Phi_r^{\tau,t_0}(x))\, r(\tau, \Phi_r^{\tau,t_0}(x)) \big\| \, \mathrm{d}\tau. \tag{E.2}$$

As a differential variant of the previous result, we state the following.

Theorem E.2 (Differentiation of a flow) *Let $\Phi_s^{t,t_0}(x_0)$ be a flow on a manifold X, defined by a vector field $v_s(t, x)$ that also depends on time and an external parameter $s \in \mathbb{R}$. Let $(s, x) \mapsto v_s(t, x) \in C_b^1$ with derivative jointly continuous in (s, t, x). Then the derivative of the flow with respect to s is given by*

$$\frac{d}{ds}\Phi_s^{t,t_0}(x_0) = \int_{t_0}^{t} \mathrm{D}\Phi_s^{t,\tau}(x(\tau))\, \frac{d}{ds} v_s(\tau, x(\tau))\, d\tau, \tag{E.3}$$

for any fixed t, t_0, and where $x(\tau) = \Phi_s^{\tau,t_0}(x_0)$.

See [DK00, Thm. B.3] for a proof of the formula for differentiation of a flow with respect to a parameter. This is a slightly modified case where the vector field is time-dependent. Theorem A.6 and Remark A.7 show that the result can be generalized to the non-autonomous case and differentiable time-dependence of v is not required.

Appendix F
Riemannian Geometry

In this appendix we recall standard facts from Riemannian geometry and establish some notational conventions. This appendix is targeted at the reader who has basic knowledge of Riemannian manifolds, but wants to have a quick refresh. For more detailed expositions see for example [GHL04, Jos08], or [Lan95] for a more abstract presentation in the context of Banach manifolds. We shall not try to be exhaustive or as general as possible in this overview.

A Riemannian manifold (M, g) is a pair of a smooth (or at least C^1, respectively C^2 for defining curvature) manifold together with a metric g: a family of positive-definite bilinear forms g_x on each tangent space $\mathrm{T}_x M$. The metric is a generalization of the Euclidean inner product on $\mathbb{R}^n$ and g_x depends in a smooth way on the point $x \in M$ in the manifold. The metric can be used to measure angles and lengths of tangent vectors, so we can define the length of a piecewise C^1 curve $\gamma : [a, b] \to M$ as

$$l(\gamma) = \int_a^b \sqrt{g_{\gamma(t)}(\gamma'(t), \gamma'(t))}\, \mathrm{d}t.$$

This length functional induces the distance function

$$d(x, y) = \inf_{\gamma} l(\gamma) \tag{F.1}$$

on M, where the infimum is taken over all piecewise C^1 curves γ connecting the points x and y. This turns M into a metric space.

Simple examples of Riemannian manifolds are $\mathbb{R}^n$ with the standard Euclidean inner product and the sphere $S^{n-1} \subset \mathbb{R}^n$ with the induced metric on its tangent bundle. Due to the Nash embedding theorem, any $C^{k \geq 3}$ Riemannian manifold can actually be realized as a submanifold of $\mathbb{R}^n$ equipped with the induced metric.

Each Riemannian manifold (M, g) has an associated linear connection, or, covariant derivative ∇ on the tangent bundle $\mathrm{T}M$. This so-called Levi-Civita connection is uniquely defined by the requirements that it is torsion-free and compatible with the metric, i.e.

J. Eldering, *Normally Hyperbolic Invariant Manifolds*, Atlantis Series in Dynamical Systems 2, DOI: 10.2991/978-94-6239-003-4,

$$\nabla_X Y - \nabla_Y X = [X, Y] \quad \text{and} \quad X\, g(Y, Z) = g(\nabla_X Y, Z) + g(Y, \nabla_X Z)$$

for all smooth vector fields X, Y, Z on M. The connection is given in local coordinates x^i by the Christoffel symbols Γ^i_{jk},

$$\nabla_{\partial_j} \partial_k = \Gamma^i_{jk}\, \partial_i,$$

where we used the Einstein summation convention for the repeated index i. The connection can be extended to the tensor bundle of M so that it satisfies the Leibniz rule.

A connection, more generally on a vector bundle $\pi : E \to M$, can also be viewed as a choice of a horizontal subbundle in $\mathrm{T}E$. There is a naturally defined vertical subbundle $\mathrm{Vert}(E) \subset \mathrm{T}E$ where $\mathrm{Vert}(E)_\xi = \mathrm{T}_\xi E_x$ for $\xi \in E_x = \pi^{-1}(x)$. A horizontal bundle $\mathrm{Hor}(E)$ is any subbundle complementary to the vertical bundle, so

$$\mathrm{T}E = \mathrm{Hor}(E) \oplus \mathrm{Vert}(E).$$

This definition of a connection is related to the definition via the covariant derivative. The horizontal bundle precisely corresponds to the tangent plane to a section s of E that is flat at a given point $x \in M$:

$$\mathrm{Hor}(E)_{s(x)} = \mathrm{Im}\big(\mathrm{D}s(x)\big) \quad \Longleftrightarrow \quad (\nabla_{\bullet} s)(x) = 0.$$

The Levi-Civita connection induces two important concepts: the geodesic flow and parallel transport. Intuitively, the geodesic flow says how to follow a straight line from an initial point along a given direction, while parallel transport defines how to keep a tangent vector fixed while carrying it along a path.[8] Both maps are defined in local coordinates as solutions of (subtly different) differential equations involving the Christoffel symbols.

The geodesic flow Υ^t is a flow on the tangent bundle $\mathrm{T}M$ and defined in local coordinates x^i by

$$\begin{aligned} \dot{x}^i &= v^i, \\ \dot{v}^i &= -\Gamma^i_{jk}(x)\, v^j\, v^k. \end{aligned} \tag{F.2}$$

Here, the v^i denote the induced additional coordinates on the tangent bundle. This geodesic flow need not be complete, that is, defined for all times. However, by the Hopf–Rinow theorem, the geodesic flow is complete if and only if M is complete as a metric space with respect to (F.1). In the following we shall assume that M is complete to simplify the exposition.

If we restrict the geodesic flow map to the tangent space $\mathrm{T}_p M$ at a fixed point $p \in M$ and to time $t = 1$, and finally project onto M, then we obtain the exponential map

---

[8] If the path is a geodesic, then parallel transport carries the initial velocity vector to the velocity vector along the entire path.

$$\exp_p = \pi \circ \Upsilon^1|_{T_pM} \colon T_pM \to M.$$

We have $D\exp_p(0_p) = \mathbb{1}_{T_pM}$, so by the inverse function theorem, $\exp_p$ is a local diffeomorphism at 0_p. The local inverse $\phi_x = \exp_x^{-1}$ of the exponential map can be viewed as a coordinate chart since $T_pM \cong \mathbb{R}^n$ isometrically. An explicit identification would require a choice of orthonormal basis in T_pM, which we shall refrain from.

Such coordinates are called normal coordinates, and locally around the point p these coordinates make M resemble $\mathbb{R}^n$ as close as possible, in the sense that the metric at p in these coordinates is equal to the Euclidean metric and the Christoffel symbols are zero. The exponential map is only a local diffeomorphism, and generally there is a maximum radius $r > 0$ such that $\exp_p \colon B(0; r) \subset T_pM \to M$ is a diffeomorphism onto its image. This is called the injectivity radius $r_{\text{inj}}(p)$ of M at the point p. The global injectivity radius of M is then defined as

$$r_{\text{inj}}(M) = \inf_{p \in M} r_{\text{inj}}(p).$$

If M is noncompact then this global injectivity radius need not be positive. The shortest path from $p \in M$ to any point x within distance $r_{\text{inj}}(p)$ is uniquely realized by one geodesic curve. In normal coordinates these curves are rays emanating from the origin. That is, let $v = \exp_p^{-1}(x)$ and $\gamma(t) = \exp_p(t\,v)$ with $t \in [0, 1]$, then $d(p, x) = l(\gamma) = \|v\|$.

Let $\gamma \colon [a, b] \to M$ be a C^1 curve, then parallel transport is a linear isometry (i.e. it preserves the metric g)

$$\Pi(\gamma) \colon T_{\gamma(a)}M \to T_{\gamma(b)}M \tag{F.3}$$

between the tangent spaces at the endpoints. We use the notation $\Pi(\gamma|_a^t)$ for parallel transport along a part of the curve. Parallel transport is defined in local coordinates x^i by the differential equation

$$\frac{d}{dt}\Pi(\gamma|_a^t)^i = -\Gamma^i_{jk}(\gamma(t))\,\gamma'(t)^j\,\Pi(\gamma|_a^t)^k \quad \text{with} \quad \Pi(\gamma|_a^a) = \mathbb{1}. \tag{F.4}$$

In (F.4) the x^i are local coordinates around the point $\gamma(t)$ with additional induced coordinates ∂_i on the tangent bundle. The representation $\Pi(\gamma|_a^t)^i$ is defined by $\Pi(\gamma|_a^t) = \Pi(\gamma|_a^t)^i\,\partial_i$. Put more abstractly, parallel transport defines a horizontal extension of a vector $v \in T_{\gamma(a)}M$ to a section of the pullback bundle $\gamma^*(TM)$, that is, a vector field $v(t)$ defined along $\gamma(t)$, which has covariant derivative zero.

On a Riemannian manifold there is the concept of curvature. A manifold is flat, i.e. it has zero curvature, if it is (locally) isometric to $\mathbb{R}^n$. The Riemann curvature R measures non-flatness on an infinitesimal level. It is given by

$$R(X, Y)\,Z = \nabla_X \nabla_Y Z - \nabla_Y \nabla_X Z - \nabla_{[X,Y]}Z,$$

which measures how much the direction of a vector Z changes when parallel transporting it around an infinitesimal loop spanned by the directions X, Y. There is a relation between the curvature and parallel transport that is important to us. If we consider holonomy, that is, parallel transport along a closed loop γ, then the deficit $\Pi(\gamma) - \mathbb{1}$ is (heuristically put) equal to the curvature form R integrated over any surface enclosed by γ. This relation can be seen as an application of Stokes' theorem and the differential statement is that the curvature R is the generator of the infinitesimal holonomy group [AS53, RW06].

References

[Ale61] V.M. Alekseev, An estimate for the perturbations of the solutions of ordinary differential equations. Vestnik Moskov. Univ. Ser. I Mat. Mech. (2), 28–36 (1961).

[AMR88] R. Abraham, J.E. Marsden, T. Ratiu, Manifolds, tensor analysis, and applications, 2nd edn, in *Applied Mathematical Sciences*, vol. 75 (Springer, New York, 1988)

[Ano69] D.V. Anosov, Geodesic flows on closed Riemann manifolds with negative curvature, in *Proceedings of the Steklov Institute of Mathematics*, No. 90, (Translated from the Russian by S. Feder, American Mathematical Society, Providence, R.I., 1967), 1969.

[APS02] B. Aulbach, C. Pötzsche, S. Siegmund, A smoothness theorem for invariant fiber bundles. J. Dyn. Differ. Equ. **14**(3), 519–547 (2002)

[AS53] W. Ambrose, I.M. Singer, A theorem on holonomy. Trans. Am. Math. Soc. **75**, 428–443 (1953)

[Att94] O. Attie, Quasi-isometry classification of some manifolds of bounded geometry. Math. Z. **216**(4), 501–527 (1994)

[Bro+09] H.W. Broer, M.C. Ciocci, H. Hanßman, A. Vanderbauwhede, Quasi-periodic stability of normally resonant tori. Physica D **238**, 309–318 (2009)

[BLZ98] P.W. Bates, K. Lu, C. Zeng, Existence and persistence of invariant manifolds for semi-flows in Banach space. Mem. Am. Math. Soc. **135**(645), viii+129 (1998).

[BLZ99] P.W. Bates, K. Lu, C. Zeng, Persistence of overflowing manifolds for semiflow. Comm. Pure Appl. Math. **52**(8), 983–1046 (1999)

[BLZ08] P.W. Bates, K. Lu, C. Zeng, Approximately invariant manifolds and global dynamics of spike states. Invent. Math. **174**(2), 355–433 (2008)

[Bre81] V.N. Brendelev, On the realization of constraints in nonholonomic mechanics. J. Appl. Math. Mech. **45**(3), 481–487 (1981)

[Car26] É. Cartan, Les groupes d'holonomie des espaces généralisés. Acta Math. **48**, 1–42 (1926)

[Cot11] É. Cotton, Sur les solutions asymptotiques des équations différentielles. Ann. Sci. École Norm. Sup. **28**(3), 473–521 (1911)

[DLS06] A. Delshams, R. de la Llave, T.M. Seara, Orbits of unbounded energy in quasi-periodic perturbations of geodesic flows. Adv. Math. **202**(1), 64–188 (2006)

[DK00] J.J. Duistermaat, J.A.C. Kolk, *Lie groups. Universitext* (Springer, Berlin, 2000)

[Dui76] J.J. Duistermaat, in "*Stable Manifolds*" Unpublished preprint, Utrecht University, Department of Mathematics (1976)

[Eic91] J. Eichhorn, The Banach manifold structure of the space of metrics on noncompact manifolds. Differ. Geom. Appl. **1**(2), 89–108 (1991)

[Fen72] N. Fenichel, Persistence and smoothness of invariant manifolds for flows. Indiana Univ. Math. J. **21**, 193–226 (1971/1972).

J. Eldering, *Normally Hyperbolic Invariant Manifolds*, Atlantis Series in Dynamical Systems 2, DOI: 10.2991/978-94-6239-003-4,

[Fen74] N. Fenichel, Asymptotic stability with rate conditions. Indiana Univ. Math. J. **23**, 1109–1137 (1973/74).

[Fen79] N. Fenichel, Geometric singular perturbation theory for ordinary differential equations. J. Differ. Equ. **31**(1), 53–98 (1979)

[GG73] M. Golubitsky, V. Guillemin, Stable mappings and their singularities, in *Graduate Texts in Mathematics*, vol. 14 (Springer, New York, 1973)

[GHL04] S. Gallot, D. Hulin, J. Lafontaine, *Riemannian Geometry. Universitext*, 3rd edn. (Springer, Berlin, 2004).

[GK02] S. Gudmundsson, E. Kappos, On the geometry of tangent bundles. Expo. Math. **20**(1), 1–41 (2002)

[Guc75] J. Guckenheimer, Isochrons and phaseless sets. J. Math. Biol. **1**(3), 259–273 (1974/75).

[Had01] J. Hadamard, Sur l'itération et les solutions asymptotiques des equations différentielles. Bull. Soc. Math. Fr. **29**, 224–228 (1901)

[Hal61] J.K. Hale, Integral manifolds of perturbed differential systems. Ann. Math. **73**(2), 496–531 (1961)

[Hal69] J.K. Hale, Ordinary differential equations, in *Pure and Applied Mathematics*, vol. XXI (Wiley-Interscience, New York, 1969)

[Has94] B. Hasselblatt, Regularity of the Anosov splitting and of horospheric foliations. Ergodic Theory Dyn. Syst. **14**(4), 645–666 (1994)

[HL06] À. Haro, R. de la Llave, Manifolds on the verge of a hyperbolicity breakdown. Chaos **16**(1), 013120–013128 (2006)

[Hen81] D. Henry, *in Geometric Theory of Semilinear Parabolic Equations, Lecture Notes in Mathematics*, vol. 840 (Springer, Berlin, 1981)

[Hil01] D. Hilbert, Ueber Flächen von constanter Gaussscher Krümmung. Trans. Amer. Math. Soc. **2**(1), 87–99 (1901)

[Hir76] M.W. Hirsch, Differential topology, in *Graduate Texts in Mathematics*, No. 33 (Springer, New York, 1976).

[Hop66] F.C. Hoppensteadt, Singular perturbations on the infinite interval. Trans. Am. Math. Soc. **123**, 521–535 (1966)

[Hör03] L. Hörmander, The analysis of linear partial differential operators, in *I, Classics in Mathematics, Distribution theory and Fourier analysis*. Reprint of the second (1990) edition (Springer, Berlin, 2003).

[HP70] M.W. Hirsch, C.C. Pugh, *Stable Manifolds and Hyperbolic Sets*, Global analysis (Proc. Sympos. Pure Math., Vol. XIV, Berkeley, California, 1968). Amer. Math. Soc., Providence, R.I., 133–163 (1970).

[HPS77] M.W. Hirsch, C.C. Pugh, M. Shub, *Invariant manifolds, Lecture Notes in Mathematics*, vol. 583 (Springer, Berlin, 1977)

[HW99] B. Hasselblatt, A. Wilkinson, Prevalence of non-Lipschitz Anosov foliations. Ergodic Theory Dyn. Syst. **19**(3), 643–656 (1999)

[Irw70] M.C. Irwin, On the stable manifold theorem. Bull. London Math. Soc. **2**, 196–198 (1970)

[Irw72] M.C. Irwin, On the smoothness of the composition map. Quart. J. Math. Oxford Ser. **23**(2), 113–133 (1972)

[Jon95] C.K.R.T. Jones, Geometric singular perturbation theory, in *Dynamical Systems* (Springer, Berlin, 1995), pp. 44–118 (Montecatini Terme, Lecture Notes in Math., vol. 1609, 1994).

[Jos08] J. Jost, *Riemannian geometry and geometric analysis. Universitext*, 5th edn. (Springer, Berlin, 2008).

[JS99] A.D. Jones, S. Shkoller, Persistence of invariant manifolds for nonlinear PDEs. Stud. Appl. Math. **102**(1), 27–67 (1999)

[Kap99] T.J. Kaper, An introduction to geometric methods and dynamical systems theory for singular perturbation problems, in *Proceedings of Symposia in Applied Mathematics, Analyzing Multiscale Phenomena using Singular Perturbation Methods*, ed. by M.D. Baltimore (American Mathematical Society, Providence, 1999), pp. 85–131

[Kar81] A.V. Karapetian, On realizing nonholonomic constraints by viscous friction forces and Celtic stones stability. J. Appl. Math. Mech. **45**(1), 42–51 (1981)

[Kli95] W.P.A. Klingenberg, Riemannian geometry, 2nd ed., in *de Gruyter Studies in Mathematics*, vol. 1 (Walter de Gruyter & Co., Berlin, 1995).

[KN90] V.V. Kozlov, A.I. Neĭshtadt, Realization of holonomic constraints. Prikl. Mat. Mekh. **54**(5), 858–861 (1990)

[Koz92] V.V. Kozlov, On the realization of constraints in dynamics. Prikl. Mat. Mekh. **56**(4), 692–698 (1992)

[Lan95] S. Lang, Differential and Riemannian manifolds, in *Graduate Texts in Mathematics*, 3rd edn, vol. 160 (Springer, New York, 1995)

[LL69] V. Lakshmikantham, S. Leela, Differential and integral inequalities: Theory and applications, vol. I, in *Ordinary Differential Equations* (Academic Press, New York, 1969), Mathematics in Science and Engineering, vol. 55-I.

[LM95] A.D. Lewis, R.M. Murray, Variational principles for constrained systems: theory and experiment. Int. J. Non-Linear Mech. **30**(6), 793–815 (1995)

[Lya07] A.M. Lyapunov, Problème général de la stabilité du mouvement. Ann. Fac. Sci. Toulouse Sci. Math. Sci. Phys. **9**(2), 203–474 (1907).

[Lya92] A.M. Lyapunov, The general problem of the stability of motion. Int. J. Control **55**(3), 521–790 (1992). Trans. by A.T. Fuller, Édouard Davaux's French translation (1907) of the 1892 Russian original.

[Mañ78] R. Mañé, Persistent manifolds are normally hyperbolic. Trans. Amer. Math. Soc. **246**, 261–283 (1978)

[Muj86] J. Mujica, Complex analysis in Banach spaces, in *North-Holland Mathematics Studies*, vol. 120 (North-Holland Publishing Co., Amsterdam, 1986). Holomorphic functions and domains of holomorphy in finite and infinite dimensions, Mathematical Notes, 107

[Pal75] J. Kenneth, Palmer, linearization near an integral manifold. J. Math. Anal. Appl. **51**, 243–255 (1975)

[Per29] O. Perron, Über Stabilität und asymptotisches Verhalten der Integrale von Differentialgleichungssystemen. Math. Z. **29**(1), 129–160 (1929)

[Per30] O. Perron, Die Stabilitätsfrage bei Differentialgleichungen. Math. Z. **32**(1), 703–728 (1930)

[Poi92] H. Poincaré, *Les méthodes nouvelles de la mécanique céleste* (Gauthier-Villars, Paris, 1892). Tome I (Les Grands Classiques).

[PS04] C. Pötzsche, S. Siegmund, C^m-smoothness of invariant fiber bundles. Topol. Methods Nonlinear Anal. **24**(1), 107–145 (2004)

[PT77] J. Palis, F. Takens, Topological equivalence of normally hyperbolic dynamical systems. Topology **16**(4), 335–345 (1977)

[PT83] J. Palis, F. Takens, Stability of parametrized families of gradient vector fields. Ann. Math. 118(2, 3), 383–421 (1983).

[Rob68] W.J. Robbin, On the existence theorem for differential equations. Proc. Amer. Math. Soc. **19**, 1005–1006 (1968)

[Roe88] J. Roe, An index theorem on open manifolds. I, II, J. Diff. Geom. **27**(1), 87–113, 115–136 (1988).

[RU57] H. Rubin, P. Ungar, Motion under a strong constraining force. Commun. Pure Appl. Math. **10**, 65–87 (1957)

[RW06] H. Reckziegel, E. Wilhelmus, How the curvature generates the holonomy of a connection in an arbitrary fibre bundle. Results Math. **49**(3–4), 339–359 (2006)

[Sak90] K. Sakamoto, Invariant manifolds in singular perturbation problems for ordinary differential equations. Proc. Roy. Soc. Edinburgh Sect. A **116**(1–2), 45–78 (1990)

[Sak94] K. Sakamoto, Smooth linearization of vector fields near invariant manifolds. Hiroshima Math. J. **24**(2), 331–355 (1994)

[Sas58] S. Sasaki, On the differential geometry of tangent bundles of Riemannian manifolds. Tôhoku Math. J. **10**(2), 338–354 (1958)

[Sch01] T. Schick, Manifolds with boundary and of bounded geometry. Math. Nachr. **223**, 103–120 (2001)

[Shu92] M.A. Shubin, Spectral theory of elliptic operators on noncompact manifolds. Astérisque 5(207), 35–108 (1992), in *Méthodes semi-classiques*, vol. 1 (Nantes, 1991).

[Šil69] L.P. Šil nikov, A certain new type of bifurcation of multidimensional dynamic systems. Dokl. Akad. Nauk SSSR **189**, 59–62 (1969).

[Tak80] F. Takens, Motion under the influence of a strong constraining force, Global theory of dynamical systems, in *Proceedings of the International Conference, Northwestern University, Evanston, Ill., Lecture Notes in Mathematics, 1979*, vol. 819 (Springer, Berlin, 1980), pp. 425–445.

[Uze+02] T. Uzer, C. Jaffé, J. Palacián, P. Yanguas, s Wiggin. The geometry of reaction dynamics. Nonlinearity **15**(4), 957–992 (2002)

[Van89] A. Vanderbauwhede, Centre manifolds, normal forms and elementary bifurcations, Dynamics reported, in *Dyn. Report. Ser. Dynam. Systems Appl.*, vol. 2 (Wiley, Chichester, 1989), pp. 89–169.

[Ver05] F. Verhulst, Methods and applications of singular perturbations, in *Texts in Applied Mathematics*, vol. 50 (Springer, New York, 2005) (Boundary layers and multiple timescale dynamics).

[Str79] S.J. van Strien, Center manifolds are not C^∞. Math. Z. **166**(2), 143–145 (1979)

[VG87] A. Vanderbauwhede, S.A. van Gils, Center manifolds and contractions on a scale of Banach spaces. J. Funct. Anal. **72**(2), 209–224 (1987)

[Whi36] H. Whitney, Differentiable manifolds. Ann. Math. **37**(2, 3), 645–680 (1936).

[Wig94] S. Wiggins, Normally hyperbolic invariant manifolds in dynamical systems, in *Applied Mathematical Sciences* ed. by G. Haller, I. Mezić, vol. 105 (Springer, New York, 1994)

[Yi93] Y. Yi, A generalized integral manifold theorem. J. Diff. Equ. **102**(1), 153–187 (1993)

[Zei86] E. Zeidler, Nonlinear functional analysis and its applications, in *Fixed-point theorems* (Springer, New York, 1986). (Translated from the German by P.R., Wadsack)

Index

J. Eldering, *Normally Hyperbolic Invariant Manifolds*, Atlantis Series in Dynamical Systems 2, DOI: 10.2991/978-94-6239-003-4,

Printed in the United States
By Bookmasters